CHALLENGING POST-CONFLICT ENVIRONMENTS

Global Security in a Changing World

Series Editor: *Professor Nana K. Poku, John Ferguson Professor, Department of Peace Studies, University of Bradford, UK*

Globalization is changing the world dramatically, and a very public debate is taking place about the form, extent and significance of these changes. At the centre of this debate lie conflicting claims about the forces and processes shaping security. As a result, notions of inequality, poverty and the cultural realm of identity politics have all surfaced alongside terrorism, environmental changes and bio-medical weapons as essential features of the contemporary global political landscape. In this sense, the debate on globalization calls for a fundamental shift from a status quo political reality to one that dislodges states as the primary referent, and instead sees states as a means and not the end to various security issues, ranging from individual security to international terrorism. More importantly, centred at the cognitive stage of thought, it is also a move towards conceiving the concept of insecurity in terms of change.

The series attempts to address this imbalance by encouraging a robust and multi-disciplinary assessment of the asymmetrical nature of globalization. Scholarship is sought from areas such as: global governance, poverty and insecurity, development, civil society, religion, terrorism and globalization.

Challenging Post-conflict Environments

Sustainable Agriculture

Edited by

ALPASLAN ÖZERDEM
*Centre for Peace and Reconciliation
Studies, Coventry University, UK*

REBECCA ROBERTS
*Centre for Peace and Reconciliation
Studies, Coventry University, UK*

Routledge
Taylor & Francis Group

LONDON AND NEW YORK

First published 2012 by Ashgate Publishing

2 Park Square, Milton Park, Abingdon, Oxon OX14 4RN
711 Third Avenue, New York, NY 10017, USA

Routledge is an imprint of the Taylor & Francis Group, an informa business

First issued in paperback 2016

British Library Cataloguing in Publication Data
Challenging post-conflict environments : sustainable agriculture. – (Global security in a changing world)
    1. Postwar reconstruction – Case studies. 2. Agriculture and state – Case studies.
    3. Sustainable agriculture – Case studies. 4. War – Environmental aspects. 5. War and society.
    I. Series II. Özerdem, Alpaslan. III. Roberts, Rebecca.
    338.1'8–dc23

Library of Congress Cataloging-in-Publication Data
Challenging post-conflict environments : sustainable agriculture / [edited] by Alpaslan Özerdem, Rebecca Roberts.
        p. cm. – (Global security in a changing world)
    Includes bibliographical references and index.
    ISBN 978–1–4094–3482–5 (hbk. : alk. paper)
    1. Sustainable agriculture – Political aspects. 2. Sustainable agriculture – Social aspects.
    3. Sustainable agriculture – Economic aspects. 4. Sustainable agriculture – Developing countries. 5. Postwar reconstruction. 6. Peace-building. 7. Civil war. 8. Ethnic conflict.
    9. War and society.
    I. Özerdem, Alpaslan. II. Roberts, Rebecca. III. Title: Sustainable agriculture.
    S494.5.S86C44 2012
    338.1–dc23   2012013756

ISBN 978-1-4094-3482-5 (hbk)
ISBN 978-1-138-27400-6 (pbk)

Contents

PART III: THE RECOVERY OF THE AGRICULTURE SECTOR

PART IV: CONCLUSION

List of Figures

List of Tables

List of Contributors

Kwadwo Asenso-Okyere is a director and member of the senior management team of the International Food Policy Research Institute (IFPRI). He holds a BSc (Honours) degree from the University of Ghana, MSc from the University of Guelph, Canada, and PhD from the University of Missouri–Columbia, USA. Before joining IFPRI he was professor and vice-chancellor of the University of Ghana and previously director of the Institute of Statistical, Social and Economic Research (ISSER) of the university. He has also worked at the World Bank, Ghana Ministry of Food and Agriculture and Ghana Cocoa Board, and served on national and international boards. He has conducted research on economic and social development and published widely. He is a fellow of the Ghana Academy of Arts and Sciences, fellow of the Association of Certified Entrepreneurs, and member of the New York Academy of Sciences.

James Bennett holds a BSc in Biological Sciences and an MSc in Ecology from the University of Warwick. He undertook his doctoral studies at Coventry University, researching small-scale cattle production systems in communal areas of Eastern Cape Province, South Africa. Since completing his PhD, he has lectured in Environmental Science at Coventry University and continued to undertake research in Eastern Cape on the management of communally held rangelands and arable land allocations as common property systems. More recently his research has focused on the sustainability of current grazing practices on former commercial farmland that has been transferred to communal ownership.

Catherine Bolten is Assistant Professor of Anthropology and Peace Studies at the University of Notre Dame. She has consulted for the United Nations World Food Programme and Physicians for Social Responsibility. Dr Bolten has conducted fieldwork in Botswana, Swaziland, Lesotho and Sierra Leone, the latter since 2003. Dr Bolten has published articles in *The Journal of Modern African Studies* and *Political Ecology*, and will publish her first book on the Sierra Leonean war for the University of California Press series in Public Anthropology, entitled *I Did It to Save My Life: Morality and Survival in Sierra Leone*, in 2012.

Sigfrido Burgos Cáceres, a national of Honduras, is an International Consultant at the Food and Agriculture Organization (FAO) of the United Nations in Rome, Italy, where he undertakes technical writing on agriculture, livestock and veterinary issues for the Animal Production and Health Division. He holds six academic degrees from four educational institutions, and has travelled widely around the world. He has published 30 peer-reviewed journal articles, five institutional working papers, numerous research and policy briefs, and is currently finishing a book on China's global resources strategy across Asia, Africa and South America. He is an avid reader of non-fiction books and speaks Spanish, English, French and Italian.

Ian Christoplos is currently a project senior researcher in Natural Resources and Poverty at the Danish Institute for International Studies. He has worked as a researcher and practitioner over the past twenty years looking at how local institutions manage risk and post-disaster/conflict recovery in rural areas. This has included research and various assignments in Africa, Asia, Latin America and the Western Balkans. His particular focus is on how an understanding of the processes under way within local institutions, such as agricultural advisory services, can guide the search for greater coherence in efforts related to risk reduction, food security and climate change and adaptation.

Marwan Darweish is a senior lecturer at the Centre for Peace and Reconciliation Studies at Coventry University, UK, and an expert in peace processes and conflict transformation. He has extensive experience across the Middle East region and a special interest in the Israeli–Palestinian conflict and ongoing peace negotiations. He has wide-ranging experience in leading and facilitating training courses and undertaking consultancies associated with conflict transformation and peace processes. He has also been involved in conflict management programmes in East and Central Africa. Previously, Dr Darweish worked as Peace and Conflict adviser at Responding to Conflict (RTC).

Brian T. Dugdill was raised on the family dairy farm in the north of England. After graduating in dairy science from the University of Reading in 1966 he has been a dairy development practitioner, first with Glaxo International and the supermarket group Asda. Since the mid-1970s, he has worked in more than 30 mainly developing countries, including Eritrea, Mongolia and North Korea. From 1976 to 1985 he led the UN 'Milk Vita' programme that established modern dairying in Bangladesh after its war of independence from Pakistan; and from 1986 to 1992 he led the multi-donor UN team that supported the rebuilding of the Ugandan dairy industry after its prolonged civil war. He presently combines the role of Chief Adviser, East Africa Dairy Development project, with food security and livestock assignments for the UN around the globe. He was awarded the UN Food and Agriculture Organization's B.R. Sen Award for 2006 and the President of Mongolia's Special Achievement Medal in 2007.

Sophal Ear is an Assistant Professor of National Security Affairs at the US Naval Postgraduate School (NPS) in Monterey, California, teaching on post-conflict reconstruction and the political economy of Asia. A TED2009 Fellow who spoke at the Oslo Freedom Forum 2010, he has just completed a book manuscript on aid dependence and democracy in Cambodia. He has contributed to the *New York*

Times, the *Wall Street Journal* and CNN.com, among other outlets. Before joining NPS, Dr Ear was a Post-Doctoral Fellow at the Maxwell School of Syracuse University. He has consulted for the World Bank and the Asian Development Bank, and was an Assistant Resident Representative for the United Nations Development Programme in East Timor during 2002–3. A graduate of Berkeley in economics, political science, and agricultural and resource economics, and of Princeton in public affairs, he moved to the United States from France as a Cambodian refugee at the age of ten.

Phil Harris is a Professor of Plant Science and Director of the Centre for Agroecology and Food Security at Coventry University. His research interests are in plant science, particularly abiotic stress physiology, tropical crop development, organic and sustainable agriculture, and agroforestry, often related to overseas development. His research and consultancy activities in sustainable agriculture and forestry have involved work in Argentina, Bangladesh, Brazil, Cape Verde, China, Cuba, Ghana, Greece, India, Jamaica, Japan, Jordan, Kenya, Libya, Nigeria, Oman, Sierra Leone, South Africa and Spain. Phil has published over 120 refereed papers, 14 books and booklets, 36 published conference papers and five edited volumes.

Moya Kneafsey is a Reader in Human Geography at the Department of Geography, Environment and Disaster Management at Coventry University. She is an experienced agro-food researcher interested in critical explorations of the relationships and values required for the 'reconnection' of consumers, producers and food, and the formulation of sustainable and ethically sound food systems. In particular, she has worked on 'alternative' food networks, local food and rural development in Europe and the Caribbean. She has published three books, and over 40 journal papers and book chapters. She is a member of Coventry University's Centre for Agro-Ecology and Food Security, and leads the working group on 'Food and Communities'.

Roy Maconachie is a lecturer in international development at the University of Bath. He undertakes field-based empirical research in sub-Saharan Africa that explores the social, political and economic aspects of food production and natural resource management, and their relationships to wider societal change. His disciplinary background is in human geography, but his work has largely been interdisciplinary in its approach, drawing principally upon anthropology and politics/political economy. He is particularly interested in the politics around natural resource management and conflict in West Africa and his current research interests focus on two main areas of enquiry: urban and peri-urban food production and resource management in northern Nigeria and Sierra Leone; and diamond mining, livelihoods and post-conflict development in Sierra Leone.

Lenard Milich has a bachelor's in meteorology and climatology, and a master's in Forestry and Environmental Studies. His doctoral studies focused on arid lands; his dissertation research across the West African Sahel examined satellite indicators of land degradation and food insecurity. He works informally with NASA, jointly testing satellite-derived areal rainfall estimates and new algorithms. He worked for seven years as food security and vulnerability analyst for the UN World Food Programme in various Asian and African countries. He started a biofuel initiative in East Africa which, in 2008, was awarded one of the World Bank's prestigious Development Marketplace awards. In early 2008 he became the Manager of Natural Resources Research at the Afghanistan Research and Evaluation Unit. He is now employed by the United Nations Office on Drugs and Crime in Afghanistan, in its Alternative Livelihoods Unit.

Liam Morgan is Programme Development Officer at Practical Action in North Darfur, Sudan. He has a professional background in manufacturing engineering and subsequently achieved an MSc in Engineering Project Management at Leeds University. Following two and a half years working with Practical Action in Khartoum on a range of development projects he returned to the UK to complete an MSc in Violence, Conflict and Development at the School of Oriental and African Studies in London. Liam has a keen interest in Sudan and the Sudanese colloquial Arabic dialect, and returned there with Practical Action at the beginning of 2011. He is currently working on the Darfur Peace and Stability Project, which allows him to pursue his interest and further his experience in promoting rural livelihoods during protracted conflict.

Gobinda Neupane has managed DFID Nepal's Community Support Programme for four years. He has an MSc in Agriculture Science from the University of London, UK, and a professional background in agricultural research and development. He worked with CARE International in Nepal for more than a decade as a manager and technical adviser for agriculture and natural resources and peacebuilding. He worked for eight years as a horticultural crop research scientist in the Pakhribas Agriculture Centre in Nepal, funded by the UK government, and two years at the Department of Agriculture in Nepal as a horticulturist. With his extensive experience in agriculture he works to improve people's livelihoods.

Alpaslan Özerdem is Director of the Centre for Peace and Reconciliation Studies, Coventry University, UK. With field research experience in Afghanistan, Bosnia-Herzegovina, El Salvador, Kosovo, Lebanon, Liberia, Philippines, Sierra Leone, Sri Lanka and Turkey, he specializes in the politics of humanitarian interventions, disaster response, security sector reform, reintegration of former combatants and post-conflict state building. He has taken an active role in the initiation and management of several advisory and applied research projects for a range of national and international organizations. He is co-author of *Disaster Management and Civil Society: Earthquake Relief in Japan, Turkey and India* (I.B.Tauris, 2006), author of

Post-war Recovery: Disarmament, Demobilization and Reintegration (I.B.Tauris, 2008), co-editor of *Participatory Research Methodologies: Development and Post-Disaster/Conflict Reconstruction* (Ashgate, 2010) and co-editor of *Child Soldiers: From Recruitment to Reintegration* (Palgrave Macmillan, 2011).

Barnaby Peacocke is head of Impact and Practice at Practical Action. He has a doctorate from work with national agricultural research systems in Africa, and international experience in humanitarian aid and the design and management of food and agriculture development programmes. He currently specializes in organizational strategy and impact, focusing on the interface between technology change and social empowerment.

Rebecca Roberts is a research fellow at the Centre for Peace and Reconciliation Studies at Coventry University, UK. She is a consultant in post-conflict reconstruction and international development with expertise in stabilization and security, governance, aid effectiveness and forced migration. She has conducted field research in Angola, Afghanistan, Cambodia, Lebanon, Sudan and South Sudan. She has advised a wide range of national and international organizations and has managed projects for the Afghanistan Research and Evaluation Unit (AREU), the Peace Research Institute Oslo (PRIO), the United Nations Institute for Disarmament Research (UNIDIR), the United Nations Development Programme (UNDP) and the Small Arms Survey, Geneva. She published *Palestinians in Lebanon: Refugees Living with Long-term Displacement* (I.B.Tauris, 2010).

Bruce A. Scholten is Honorary Research Fellow in the Department of Geography, Durham University, and lectures at Sunderland University Faculty of Applied Sciences. His research interests include cooperative dairy farming, animal welfare in conventional and organic farming, and transport in rural health. He has written on agricultural policy for a variety of international publications and his current research focus is the political economy of food. In 2010 he published *India's White Revolution: Operation Flood, Food Aid and Development*. He grew up on a dairy farm near Lynden, Washington, USA.

Richard Slade is a PhD researcher at the Centre for Peace and Reconciliation Studies at Coventry University. His research topic is faith and secular peacebuilding in the UK, applied in areas where there is identity-based intercommunity tension and hostility involving right-wing extremism. His research builds on an MA in Peace Studies from the University of Bradford and a career of 40 years in social work, social care and health care as a practitioner, manager and consultant focusing on child protection, mental ill health and partnership working.

Sylvia Stoll is a Marie Curie Fellow at the Centro de Ciencias Humanas y Sociales, Consejo Superior de Investigaciones Científicas (Spanish National Research Council), in Madrid, and a PhD candidate at the Institute for Media Studies and at the Institute for International Law of Peace and Armed Conflict, Ruhr-Universität Bochum, Germany. She has studied Anthropology, Media Arts and Humanitarian Assistance and worked as a link/coordinator for the NOHA Master's Programme at Ruhr-Universität Bochum before accepting her fellowship in the SPBUILD Initial Training Network on Sustainable Peacebuilding. During an Erasmus Mundus exchange at Universitas Gadjah Mada in Yogyakarta, Indonesia, in 2008 she conducted field research in Aceh.

Sindu Workneh Kebede is a doctoral student at Humboldt University, hosted and sponsored by the German Institute for Economic Research (DIW Berlin, Germany). She has an MSc in Economic Policy Analysis and a BA degree in Economics from Addis Ababa University, Ethiopia. Sindu worked at the International Food Policy Research Institute (IFPRI) and obtained extensive research experience. At IFPRI she was actively engaged in developing the strategic and operational plan of the Sierra Leone Agricultural Research Institute (SLARI) under the project 'Revamping agricultural research in Sierra Leone', and involved in the project 'Revitalizing agricultural research in Nigeria'. She has trained in advanced research methods for micro-level analysis of conflict and is a member of MICROCON and Ethiopian Economics Association.

Lionel Weerakoon researches conservation farming, agroforestry for upland crops and nitrogen-fixing plants to improve crops. He has served in the Sri Lanka Department of Agriculture and the International Water Management Institute on the Watershed Development Project. He was the first Executive Director of the Wildlife Trust of Sri Lanka and a senior lecturer on sustainable agriculture at the University of Ruhuna. Dr Weerakoon is a Founder Director of the Sewa Lanka Foundation, a member of the National Committee of the Movement for National Land and Agricultural Reform (MONLAR), and the Chief Adviser to the Center for Sustainable Agriculture and Development. At present, he is the Chairman of the Thematic Committee on Alternative Agriculture for Self Reliance of the National Science Foundation. He represents the South Asian Region for the World Commission on Sustainable Agriculture of La Via Campesina.

Julia Wright has 25 years of professional research and development experience in organic farming, food security and rural livelihoods worldwide, she has specialized in regenerating land and livelihoods in vulnerable and disaster contexts. She undertook her PhD research at Wageningen University on the impact of the loss of food and fuel supplies on the farming and food system in Cuba in the 1990s. She is currently Deputy Director of the Centre for Agroecology and Food Security, a joint initiative of Coventry University and Garden Organic. She is also a Deployable Civilian Expert with the Stabilization Unit of DFID/FCO/MOD.

Acknowledgements

We wish wholeheartedly to thank the contributors to this book. We are deeply grateful for their professionalism, commitment, perseverance and scholarship that they brought to bear in producing this volume.

List of Abbreviations

AFRC	Armed Forces Revolutionary Council (Sierra Leone)
BiH	Bosnia-Herzegovina
CBO	community-based organization
CLARA	Communal Land Rights Act (South Africa)
CPA	Communal Property Association (South Africa)
CPI	Common Property Institution (South Africa)
CSRS	Center for Stabilization and Reconstruction Studies (USA)
DACAAR	Danish Committee for Aid to Afghan Refugees
DANIDA	Danish International Development Agency
DDR	Disarmament, Demobilization and Reintegration
DFID	Department for International Development (UK)
DLA	Department of Land Affairs (South Africa)
DoA	Department of Agriculture (South Africa)
EADD	East Africa Dairy Development
ECDA	Eastern Cape Department of Agriculture (South Africa)
FAO	Food and Agriculture Organization of the United Nations
FUPAP	Freetown Urban and Peri-urban Agriculture Platform (Sierra Leone)
GAM	Gerakan Aceh Merdeka/Free Aceh Movement (Aceh)
GHI	Global Hunger Index
IAR	Institute of Agricultural Research (Sierra Leone)
ICRAF	International Centre for Research in Agroforestry/World Agroforestry Centre
ICRC	International Committee of the Red Cross/Crescent
IDP	internally displaced person
IFPRI	International Food Policy Research Institute
ILRI	International Livestock Research Institute
INGO	international non-governmental organization
IOM	International Organization for Migration
IPCC	Intergovernmental Panel on Climate Change
LARP	Land and Agrarian Reform Project (South Africa)
LTTE	Liberation Tigers of Tamil Eelam (Sri Lanka)
M&E	monitoring and evaluation
MAAIF	Ministry of Agriculture, Animal Industry and Fisheries (Uganda)
MAFFS	Ministry of Agriculture, Forestry and Food Security (Sierra Leone)
MFMR	Ministry of Fisheries and Marine Resources (Sierra Leone)
MONLAR	Movement for Land and Agricultural Reform (Sri Lanka)

NCDDR	National Commission for Disarmament, Demobilization and Reintegration (Sierra Leone)
NDDB	National Dairy Development Board (India)
NWGA	National Wool Growers' Association (South Africa)
OPT	Occupied Palestinian Territories
PADO	Polong Agricultural Development Organization (Sierra Leone)
PAPD	Participatory Action Plan Development
PNA	Palestinian National Authority
PRSP	Poverty Reduction Strategy Paper
RA	Residents' Association
RDN	El Fasher Rural Development Network (Darfur)
RUF	Revolutionary United Front (Sierra Leone)
UNDP	United Nations Development Programme
UNODC	United Nations Office on Drugs and Crime
UPA	urban and peri-urban agriculture
USAID	United States Agency for International Development
VDC	Village Development Committee (Darfur)
VFA	Vukani Farmers' Association (South Africa)
VNRHD	Voluntary Network for Rural Help and Development (Darfur)
WDA	Women's Development Association (Darfur)
WFP	World Food Programme
WGA	Wool Growers' Association (South Africa)
WTO	World Trade Organization

Introduction

This book conceptualizes the challenges of developing sustainable agriculture in post-conflict environments as well as identifying practical solutions to achieve sustainable agricultural production, which is central to the survival of humanity. However, the environment, infrastructure, social structures and skills, economy, and governance systems necessary to maintain agricultural production are fragile and easily damaged during conflict. Without sustainable agriculture, populations remain vulnerable, increasing the likelihood of a return to conflict. Therefore, sustainable agriculture is central to effective post-conflict recovery that provides human security as well as stability and rule of law.

In fact, the theme of 'Stabilization Agriculture' is a joint area of research focus between the two flagship centres of Coventry University: Centre for Peace and Reconciliation Studies (CPRS) and Centre for Agroecology and Food Security (CAFS). To address the need for comprehensive research on the challenges and recovery of agriculture in challenging environments such as disaster and conflict affected environments the 'Stabilization Agriculture' initiative has already taken a number of key activities such as the organization of 'The Stabilization Agriculture Workshop' at the Coventry University London Campus in June 2011. The event was attended by a wide range of delegates, including academics, researchers, consultants, and representatives from non-governmental organizations, and played a key role in the development of this edited volume. One of the key findings of the Workshop was that despite the importance of sustainable agriculture in such challenging environments, little has been published to date on the subject.

Therefore, to address the paucity of literature on this subject effectively the book covers issues of local and global concern ranging from community-level subsistence farming to industrialized farming, and from localized shortages of natural resources to climate change. By adopting a holistic multidisciplinary approach which identifies key themes and case studies, this book sets the scene for the debates surrounding sustainable agriculture in post-conflict environments.

Overall, this book brings together two sets of literature and practice that are often considered separately. The first set of literature is broadly within the sphere of post-conflict studies, while the second relates to agriculture, development and sustainability. Although violent conflict takes place in societies that have predominantly agricultural economies, issues connected with agriculture are often dealt with (in the policy and academic worlds) as though they are apolitical and somehow separate from wider political dynamics. Therefore, the book offers originality in terms of its concept in that it sees 'fixing' agriculture as more than merely a technical matter and instead one with social, political and cultural consequences as well as the obvious economic ones. Overall, there is originality in the interdisciplinary nature of the book. Interdisciplinary often means bringing together a political scientist and a sociologist, but in this case it means bringing together natural and social scientists, as well as those with practical experience in development and agricultural contexts.

The key questions responded by a set of multidisciplinary contributions in this volume can be categorized in three main groups. First, the book focuses on those questions in relation to strategic challenges and priorities such as: governance in post-conflict countries is usually weak so how are agricultural policies initiated? How are national priorities reconciled with the international priorities? Are policies driven by the international community or the national government? The chapters also shed light on who implements these policies? What are the tensions between national policies and local community led initiatives? At what level do external actors decide to act – do they support national or local strategies?

Second, the book focuses on a number of key areas of challenges for the recovery of agriculture in challenging environments. For the wider economic context, the book deals with such questions as how do post-conflict countries compete in international trade? What are the costs to sustainable agriculture and to the food security of the population if countries engage in international trade? What are the driving ideologies, private market led economic policies or centralized public funding from the national government? On the other hand, in terms of environmental impacts, natural resources are damaged directly through conflict because areas are bombed or deliberately polluted for example, and indirectly because populations are forced to adopt environmentally damaging practices. After the conflict, areas can remain contaminated by explosive remnants of war, which forces populations to take risks or over-farm accessible areas. Therefore, one of the key questions for the volume is how can environmental damage be managed to encourage sustainable agriculture? Moreover, the physical and mental trauma caused by conflict reduces the size and effectiveness of the work force. At the same time, specialized local knowledge, which has been developed over generations, is lost because people are killed or move away. In other words, how can the labour force be rehabilitated to work effectively, retaining the best traditional practices post-conflict?

The third set of questions are primarily related to the recovery of the agriculture sector, which can be divided into a number of sub-groups and a wide range of case studies from around the world have been used to illustrate how these issues can be addressed. The following themes and questions are covered:

Land-related Issues

Establishing landownership and access rights following a conflict is very sensitive, particularly as disputes over land rights may have played an instrumental role in the conflict. How are rights and ownership of land balanced against the needs to develop sustainable agricultural production? How can technically practical solutions be implemented that are satisfactory to conflict parties and do not risk a return to conflict?

Water and Irrigation

Disputes over water can be at the root of a conflict, and irrigation systems can be damaged during conflict. How can systems that allow equitable access to water and provide effective irrigation be developed to ensure sustainable agriculture as well as stability and security?

Farming Techniques and Know-how

Traditional farming techniques and know-how can be lost during conflict? How can the knowledge be recovered? How can traditional practices be reconciled with imported expertise? Who decides which approaches are appropriate?

Financial Support

What type of financial or practical support is needed? What mechanisms (through individuals, organizations, grants, micro-credit, etc.) can be used to deliver financial assistance?

Livelihoods and Income Generation

What investments are needed post-conflict to develop local markets for agricultural produce – irrigation, transport infrastructure, market places, processing plans?

With these questions in mind the book is divided into four main parts. In Part I, the introductory chapters by Moya Kneafsey (*Global Agriculture and the Challenge of Sustainability*), and Alpaslan Özerdem and Rebecca Roberts (*The Impact of Conflict on Agriculture and Post-conflict Reconstruction Challenges*) set the parameters of the debates surrounding sustainable agriculture and food security, the impact conflict has on agriculture through damage to social, economic, environmental and governance structures, and the challenges to the post-conflict reconstruction of war-affected populations and the agricultural sector. These chapters demonstrate the challenges to developing sustainable agriculture systems and food security in stable conditions and highlight the vulnerability of sustainable agriculture to violent conflict. The final chapter in Part I by Phil Harris and Alpaslan Özerdem (*Projections on Future Challenges: Climate Change*) focuses on climate change, highlighting how such an emerging issue could lead to future conflicts and what approaches can be taken to adapt agriculture in post-conflict environments.

Part II contains four case studies and looks at the immediate and long-term impact of conflict on agriculture. Kwadwo Asenso-Okyere and Sindu Workneh Kebede (*The Effects of Civil War on Agricultural Development and Rural*

Livelihood in Sierra Leone) look at the effects of the civil war of Sierra Leone on agricultural development and rural livelihood and the prospects of economic growth for the future. Using Nepal as a case study, Gobinda Neupane and Richard Slade (*How Conflict Affected Agriculture in Nepal*) identify the many different ways in which conflict has affected the country's agricultural sector. The chapters by Rebecca Roberts (*The Legacy of War: Unexploded Cluster Munitions in Southern Lebanon*) and Sylvia Stoll (*Women in Aceh: Conflict and Changing Agricultural Roles*) examine the lasting legacy of conflict by looking at cluster munition contamination in southern Lebanon and the changing agricultural roles of women in Aceh.

The third section uses case studies from around the world to examine ways in which agriculture continues during conflict and can be rehabilitated post-conflict. Julia Wright and Lionel Weerakoon (*Taking an Agroecological Approach to Recovery: Is It Worth It and Is It Possible?*) and James Bennett (*Post-apartheid Struggles: Land Rights and Smallholder Agriculture in South Africa*) look at the effects in South Africa, Afghanistan and Sri Lanka that resources such as land, seeds and access to markets have on the local population and agricultural production.

Sigfrido Burgos Cáceres and Sophal Ear (*Cambodia: The Challenge of Adding Value to Agriculture after Conflict*), and Bruce A. Scholten and Brian T. Dugdill (*Avoiding Dairy Aid Traps: The Cases of Uganda, India and Bangladesh*) examine national-level strategies for developing the contribution agriculture can make to the economy in a number of different contexts.

Chapters by Lenard Milich (*Explicitly Licit: Stemming the Sand Tide in Kohsan District, Herat Province, Afghanistan*), Marwan Darweish (*Olive Trees: Livelihoods and Resistance*), and Liam Morgan and Barnaby Peacocke (*Practical Action in North Darfur*) point out that agricultural production in Afghanistan, Palestine and Darfur continues during conflict and that carefully tailored external support can have a positive impact, despite the violence.

Using examples from Central Asia, Afghanistan and Bosnia-Herzegovina, Ian Christoplos (*Agricultural Information amid Conflict: for Whom and for What?*) also talks about the importance of accurate information to inform the design of agriculture programmes in post-conflict environments. Roy Maconachie (*Youth, Associations and Urban Food Security in Post-war Sierra Leone*) and Catherine Bolten (*The Only Way to Produce Food is to Cooperate and Reconcile?*) approach agriculture in Sierra Leone from different perspectives, examining the growth of urban agriculture and the dangers of linking food security and peacebuilding initiatives respectively.

The final section summarizes the lessons learned from the various case studies and highlights gaps in research. It argues that to date, despite the need for food security to ensure the continued survival of the human race, there has been little attention to how agriculture could and should be rehabilitated post-conflict.

PART I
Concepts, Issues and Challenges

Chapter 1

Global Agriculture and the Challenge of Sustainability

Moya Kneafsey

Introduction

It may seem obvious to state that agriculture is essential to human survival, but what is not so obvious is how to organize this most central of all productive activities in ways that are 'sustainable'. Even less clear is how this can be achieved in post-conflict environments. People living and working in such environments are faced with particular and immediate difficulties in rebuilding agricultural systems, including damage to ecological, physical and social infrastructure. In addition, they are faced with the same underlying ethical and ideological issues that concern those attempting to shape sustainable agriculture in many regions throughout the world, not just those in post-conflict situations. Within this context, the aim of this chapter is to provide an overview of the key debates about what 'sustainable agriculture' means and how it should be practised. It is argued that attempts to construct sustainable agriculture in post-conflict environments are set against the backdrop of an unequal and imperfect global agro-food system in which the meaning of 'sustainability' has long been contested. The complications inherent in post-conflict situations are often layered on top of existing structural inequalities and unsustainable agricultural practices, which in some cases may have contributed to the cause of conflict in the first place. The challenge of imagining and implementing sustainable agriculture in post-conflict situations is thus magnified to daunting proportions. The chapter begins with a review of the current context for discussions about agricultural sustainability, described here in terms of a new food (in)security. It then provides an overview of the conflicting world views which have shaped discussions about how to practise sustainable agriculture, before offering some brief conclusions.

The Challenges to Contemporary Agriculture

Current debates about sustainable agriculture are often located with reference to the challenge of ensuring global food security. The concept of 'food security', a term once used primarily with reference to developing countries (Jarosz 2011), is now being used in relation to developed countries too (MacMillan and Dowler

2011). A full review of the different meanings of food security is beyond the scope of this chapter, but in general it is understood to exist when 'all people, at all times, have physical, social and economic access to sufficient, safe and nutritious food which meets their dietary needs and food preferences for an active and healthy life ...' (FAO 2003: 29). Recent concerns about food security have been prompted by the increases in the number of hungry people between 1995–7 and 2004–6 and the high food prices and economic downturn that started in 2008 (World Food Programme 2011, FAO 2010). The food price spike of 2008 provoked riots in several regions, and the continued rises in food and energy prices are having an increasing impact on consumers in developed countries too (Dowler et al. in press).

Not only have the rising food prices cast doubt on the realism of ever achieving the Millennium Development Goal of halving the number of people who suffer from hunger by 2015, but the triple challenges of population growth, resource shortages (for example, energy and water) and climate change have raised the spectre of ever-increasing numbers of hungry people. This has prompted the international policy community to recognize that agricultural development is not just a 'mundane' issue of production systems, but is 'a core component of national and international security, global health, natural resource management, and climate change' (Naylor 2011: 234). Scientists and campaigners alike have argued that unless rapid improvements in agricultural productivity are achieved in tandem with efforts to stabilize global geopolitical insecurities and tackle poverty, the consequences for humanity will be dire (Beddington 2010). Moreover, and to add complexity to the problem, improvements in agricultural yield have to be undertaken without replicating the industrialized farming methods which have contributed to a number of serious environmental problems. The recent Foresight report (2011), commissioned by the UK government, lists these as: soil loss due to erosion, loss of soil fertility, salination, water being extracted faster than it can be replenished, overfishing, heavy reliance on fossil fuel energy for synthesis of nitrogen fertilizers and pesticides, emissions of greenhouse gases and other pollutants into the environment. The report urges policymakers to consider the *global* food system, from farm to plate, and argues that there is need for radical change. It recognizes five key challenges – balancing future demand and supply sustainably, to ensure that food supplies are affordable; ensuring there is adequate stability in food supplies, and protecting the most vulnerable when volatility does occur; achieving global access to food and ending hunger (recognizing that producing enough food so that everyone *potentially* can be fed is not the same thing as ensuring food security for all); managing the contribution of the food system to the mitigation of climate change; and maintaining biodiversity and ecosystem services whilst feeding the world.

In order to ensure food security, sustainable agriculture and food systems are required. The exact form that such systems should adopt is not agreed though, and it is often argued that there is no single blueprint for sustainability (Hinrichs 2010). Indeed, many have recognized that sustainability is a vague concept, whose meaning depends on political and ideological perspective (Koç 2010, Robinson

2009). The general understanding, famously drafted in the 1987 Brundtland Report, is that 'sustainable development' meets the needs of the present, without compromising the ability of future generations to meet their own needs. There is much debate about how such development is to be achieved, what form it should take, and whose needs are to be met. There are many who argue that there are inherent and irreconcilable tensions between 'sustainable' and 'development', especially when the latter is predicated on economic growth models that depend on the continued exploitation of finite natural resources and that have historically favoured countries in the global North (Blewitt 2008, Redclift 2005). In terms of defining 'sustainable agriculture', it is clear that there is regional differentiation in the types of challenges to be faced, as well as major differences between agriculture in developed and developing countries. Developed-world agriculture is dominated mainly by production systems reliant on non-renewable fossil fuels and intensive fertilizer and pesticide use, and is also characterized by consumption patterns which lead to high levels of food waste and diet-related ill health. As Robinson (2009) suggests, it is in the developed world where the greatest changes are required to move to more sustainable and less climate- and world-damaging practices. Yet whilst the developed world causes most damage through its carbon emissions and wasteful food systems, developing countries have also adopted the industrialized agricultural model in order to supply export markets, often with devastating environmental and social consequences. Naylor (2011), for example, illustrates that Brazil and Indonesia have been converting large amounts of forestland into food and feed production, especially for soybeans, oil palm and cattle – much of which is destined for markets in temperate and subtropical regions. In both cases, smallholders have been marginalized, and forest clearance is enabled because land institutions are inadequate and poor governance has allowed the spread of fraudulent land titles, regional land grabs and widespread violence and conflict.

Some developing countries have also followed a 'nutrition transition' (Popkin 2003) characterized by increased consumption of 'Westernized' diets high in saturated fats, salt and sugar, to the extent that obesity amongst the new middle classes in countries such as India and Mexico is becoming a serious health concern. At the same time, it is developing countries that bear the brunt of hunger and malnutrition (Bailey 2011). They are also the home of the majority of the world's 1.5 billion resource-poor smallholders (CSRS 2010), and much of the world's food is produced in them. Given this, it is striking and significant that so many of the world's hungry are actually farmers, who struggle to sustain themselves and their families due to a range of well-known reasons summarized by Naylor (2011), as follows: lack of inputs such as high-yielding seed varieties, fertilizers and credit; the inability to pay for such inputs; an absence of irrigation; shrinking per capita farm size with population growth; underdeveloped markets and infrastructure such as roads; poor education and health and a shortage of agricultural extension and other services. Thus, to paraphrase Naylor (2011: 243), many families find themselves in a 'poverty trap' in which they produce low-yielding, low-valued staples for home consumption and limited markets during a short rainy season. Due to lack of storage facilities, the harvest surplus has to be sold immediately, which drives the short-term market price down. With low prices and poor market connectivity there is little incentive to adopt new technologies that might increase productivity and enhance economic development. Small-scale farmers in developing regions are also most at risk of reduced agricultural productivity and hunger due to conflict. Recently, farmers – especially in sub-Saharan Africa – are also threatened by the increasing number of land grabs by foreign governments and asset management firms, many of which are currently leaving large tracts of potentially productive agricultural land fallow (Naylor 2011). On top of all this, farmers in poor countries are most vulnerable to climate change, as temperatures are forecast to rise most in tropical and subtropical areas, and water-scarce regions will face even drier conditions (Huang, von Lampe and van Tongeren 2011).

Hinrichs (2010) notes that there has been a tendency to focus on the environmental dimensions of agricultural practice, such as attempts to make farm production more energy-efficient and less destructive of natural resources. However, recognizing that sustainability has social and economic as well as environmental dimensions, more recent thinking has emphasized aspects such as social equity, justice and governance, and has also tended to recognize the interconnections of agricultural production with 'systems' or 'networks' of consumption (for instance, Goodman and Du Puis 2002). It is becoming increasingly clear that sustainable agriculture is about so much more than the techniques and knowledge required to produce food without degrading natural resources. As noted by Watson (2011), many of the technologies and practices needed to meet the challenge of sustainable agriculture already exist, including knowledge about more effective soil and water management, microbiological techniques to suppress diseases in soils, and conventional plant breeding to improve crop varieties. Naylor (2011) also provides clear recommendations on some of the requirements of sustainable agriculture in developing countries, including small-scale irrigation systems and the use of mixed farming systems. Others have considered the scale at which different farming systems should be enacted, recognizing that to conserve environmental resources, a combination of extensive and intensive systems may be required. Thus Benton et al. (2011) propose that some land can be farmed extensively over a large area, thereby producing less food but more ecosystem services on the same land (known as 'land sharing'). This would be suitable for regions with small fields, steep valleys or large amounts of non-cropped habitat which are unsuitable for intensive production, but are high in biodiversity. Other land can be farmed intensively over a smaller area, using inputs, labour and capital to increase outputs per unit of space, and the remaining land can be saved to be managed for ecosystem services ('land sparing'). This, they suggest, would be more suitable for flat landscapes with fertile soil, naturally lower in biodiversity.

Although there is much knowledge about sustainable farming systems and techniques, new problems such as climate change and emerging animal diseases require new solutions (Watson 2011). To address these problems, using existing and new knowledge, many observers have noted that there is an urgent need to invest in agricultural research, and to develop effective institutional structures which can manage common property resources, globally and locally. Effective national governance structures are required to protect land rights and ensure access to credit, water, technology and seeds for small farmers. International governance should promote fairer trade for developing countries, such as the removal of trade barriers to products for which they have competitive advantage. All of this needs to be done with the active involvement and participation of small farmers and peasants who depend directly upon agricultural sustainability for their livelihoods (Bailey 2011, Dixon and Gulliver with Gibbon 2001, Robinson 2009). This in turn means that achieving sustainable agriculture requires attention not only to the productive capabilities of farmers and environment, but also to the ways in which agriculture is embedded in global structures of poverty and inequality as well as more localized struggles for resource control. Because of agriculture's centrality to livelihoods and ecosystems, as well as to mitigating climate change, controversies over how to implement agricultural sustainability become embroiled in wider debates over the best ways of organizing societies and resources at different scales. It is to these wider debates that the chapter now turns in order to understand the competing agendas that are shaping current agricultural sustainability discourse.

Operationalizing Sustainable Agriculture: Conflicting World Views

Whilst there is much discussion about techniques for sustainable agriculture, there is little agreement on how it should be operationalized. This is partly because there are many different geopolitical contexts and farming systems, but it is also because there are profoundly different perspectives on the relationship between society and environment. The debate is often characterized as a contest between technocentric and ecocentric approaches (Robinson 2009), although critics have suggested that this dualistic analysis belies the complexity of the debate and the elements of overlap between seemingly contrasting discourses and practices (Hinrichs 2010). For the sake of clarity, this chapter sketches out two broad – but by no means homogeneous – world views, which are characterized here as 'neoliberal' and 'oppositional'. In an attempt to avoid oversimplifying the debates, reference is made to the points where these two world views overlap. Neoliberalism seeks to achieve sustainability through the marketization of environmental resources and human relationships in the pursuit of profit maximization. In this system, profits accumulate through extended value chains which are dominated by transnational corporations (TNCs) which export agricultural commodities such as wheat, soybeans, palm oil and high-value-added crops such as green beans, baby sweetcorn and mangetout, predominantly for Northern consumers (Barrett, Browne and Ilbery 2004). Through consolidation in the private sector, a small number of giant TNCs in agribusiness, fisheries, processing and retail now dominate the global agro-food system (Hendrickson and Heffernan 2002) at the expense of small-scale producers in the global South and North.

Neoliberal discourses vary according to geographical context (Busch 2010) but in general, sustainability is used by neoliberal actors as a 'legitimizing ideology to justify commodification, industrialization and globalization of the agri-food system' (Koç 2010: 38). Powerful biotechnology companies, as key beneficiaries of neoliberal policies, position themselves as offering solutions to the problem of food security by patenting seeds and technologies which will, they say, lead to greater and more reliable crop productivity. Their interest is served by framing food security primarily in terms of ensuring that there is enough food supply to feed the world's growing population – when it is clear that feeding people is not just about how much food is available; if it were, millions of people would not be hungry today. An obvious example is Monsanto (2011), describing itself as 'one of the world's leading companies focused on sustainable agriculture'. Monsanto's definition of sustainable agriculture is summarized as 'Produce more, conserve more, improve lives' and this will be achieved through advanced plant breeding, biotechnology and improved farm management to produce better seeds, faster-growth varieties and drought-resistant varieties. For example, YieldGard® Rootworm technology has been developed to establish better-rooted maize plants that use soil moisture more effectively. Whilst it could be argued that these innovations have an important place in sustainable agriculture, critics point out that large biotechnology companies such as Monsanto have benefited from the World Trade Organization's global legal regime, which regulates and promotes the extension of markets globally, but has no jurisdiction in relation to labour, the environment or human rights (Busch 2010). For example, the World Intellectual Property Organization was developed alongside the WTO to ensure that intellectual property rules would be globally harmonized (Busch 2010). This means that seeds can now be regarded as the intellectual property of a company rather than as a common-pool resource. According to farming campaign groups, this has disempowered small farmers, who now depend on multinational companies to provide seeds and other inputs, instead of conserving their own seeds and farming knowledge-practices. The case of Monsanto provides an illustration of how the distinctions between 'neoliberal' and 'oppositional' discourse are often blurred, with the latter often being appropriated by powerful neoliberal interests to construct a more favourable public image. Thus Monsanto's web pages stress that they want to work *with* farmers to make agriculture more sustainable, and in 2006 the company adopted its own Human Rights Policy.

In addition to intellectual property laws, production subsidies and trade restrictions in the USA and EU have also benefited larger farms and businesses in the global North, enabling them to sell commodities at below-production costs. As McMichael (2011) notes, WTO governance includes lowering trade barriers, eliminating Southern farm protection and exposing peasants to a crisis of low prices – 'Northern farmers and Southern peasants alike have been displaced by a price system privileging soil and water-depleting mono-cultures.' Even though it is still the case that the majority of agricultural production is undertaken by small-scale or peasant farmers (ETC 2009), the trend towards increasingly globalized, industrialized food systems supported by international trade and intellectual property rights that favour large corporations has provoked dismay amongst those concerned for the fate of farming communities, local economies and ecosystems. They have argued that industrialized agriculture controlled by international agribusinesses supplying giant retail corporations and propped up by powerful global institutions has resulted in the marginalization of peasant farmers, the loss of traditional knowledge-practices, rural depopulation, an overdependence on agricultural exports, deforestation and habitat loss leading to species extinction and the degradation of soils and water sources (Weis 2007).

National governments and international financial institutions generally seem keen to emphasize neoliberal solutions which combine technological innovation with 'efficient'-functioning markets to the problem of agricultural sustainability. For example, the Foresight report (2011) commissioned by the UK government (although not official policy) argues that 'sustainable intensification' is required to raise yields, increase the efficiency with which inputs are used and reduce the negative effects of food production. It is based on the principle of sustainability whereby resources are used at rates that do not exceed the capacity of the earth to replace them. The report stresses the need for evidence-based policy decisions and investment in research into new varieties of crops and livestock, as well as the preservation of rare breeds and wild relatives of domesticated species. Also needed is research into nutrition, soil science, animal science, agroecology, agricultural engineering and aquaculture. It argues that market and environmental incentives must be aligned in order to support environmentally sustainable practices. The report contains much that 'oppositional' campaigners would agree with, such as the reference to agroecology and land rights. However, and in contrast to many 'oppositional' voices, it firmly rejects food self-sufficiency as a solution to food security. Instead, it argues that the benefits of globalization must be maximized and distributed fairly, partly through the revitalization of agricultural extension services to increase the skills and knowledge of food producers, who are often women.

In contrast to neoliberal concepts of sustainability, 'oppositional' discourses often propose no- or low-growth scenarios for human development, and are often framed in terms of 'resistance' to globalization and corporate-driven agricultural development. Such discourses incorporate the views of diverse groups from the global South and North. La Via Campesina is a prominent actor in this loose alliance of alternative voices; the Slow Food Movement, various organic farming organizations and local food movements can also be included. This is not to say that a homogeneous voice or movement has formed, or that one single solution is being proposed – on the contrary, a hallmark of many of the groups and affiliations involved is that they embrace, celebrate and recognize diversity as a key component of their visions for sustainable agriculture. They recognize that one-size solutions will not work, and that sustainable agriculture will be attuned to the environmental, social and cultural conditions of different places. A range of farming systems can be included under this umbrella term, including organic, biodynamic, permaculture, low-input, resource-conserving and regenerative systems (Robinson 2009). However, what many of them share is what McMichael describes as an 'epistemic orientation' which rejects the reduction of ecological processes to marketable forms.

An illustration of the different epistemic orientations within neoliberal and 'oppositional' discourse can be found in the ways in which the concept of multifunctional agriculture is deployed. In general terms, multifunctionality values farming not only for its product output, but for its contribution to ecosystem management, landscape maintenance, rural employment, and preservation of local knowledge, cuisines and heritage (IAASTD 2009, McMichael 2011). Agriculture is thus presented as a public good. In world trade agreements, which help to sustain neoliberal economies, the principles underlying multifunctionality have been reduced to an 'environmental governance function within the logic of a neoliberal market calculus' (McMichael 2011). Farmers in the global North are paid directly for environmental services, which allow agricultural commodities to be sold at below-production-cost prices. This continues to work against producers in the global South. Moreover, as payments match farm size, the biggest farms have benefited most and small and medium-sized farms in the global North have been disadvantaged.

McMichael argues that the 'food sovereignty' movement represents a radically different version of multifunctionality, and a real critique of the neoliberal project. La Via Campesina, one of the most prominent food sovereignty campaign groups, shares multifunctional agriculture as a cornerstone, but seeks to achieve this through revaluing farming as the occupation of the majority of the world's population, positioning farmers as guardians of the land and regarding land as a public good. Advocates of food sovereignty see food self-reliance at community or national level as the foundation not just for sustainable agriculture but for a sustainable global community. They recognize that smallholdings are still the major source of staple foods and proven to be as or more productive and resilient than soil-depleting industrial monocultures (Altieri 2008, cited in McMichael 2011). In addition, a key aspect of many of these progressive discourses of sustainability is that democratic participation in food systems is regarded as an essential element. This is distinct from – and more far-reaching than – the calls for 'good governance' that appear in much neoliberal policy rhetoric. As summarized by La Via Campesina (2011), food sovereignty prioritizes:

> the right of peoples to healthy and culturally appropriate food produced through sustainable methods and their right to define their own food and agriculture systems. It develops a model of small scale sustainable production benefiting communities and their environment. It puts the aspirations, needs and livelihoods of those who produce, distribute and consume food at the heart of food systems and policies rather than the demands of markets and corporations.

Interestingly, in another illustration of the blurring of boundaries between neoliberalism and its critics, the IAASTD (International Assessment of Agricultural

Knowledge, Science and Technology for Development) (2009: 3) report, representing the outcome of a global consultative process initiated by the World Bank and the Food and Agriculture Organization in 2002, is clear that 'business as usual is no longer an option' in a world of 'asymmetric development, unsustainable natural resource use and continued rural and urban poverty'. It recognizes that the impressive scientific and technological achievements which led to increased productivity during the latter half of the twentieth century have also resulted in unintended social and environmental consequences. It emphasizes empowering small-scale farmers through legal frameworks to ensure access to land, recourse to fair conflict resolution and progressive evolution of intellectual property rights regimes. It also argues for policies to support the revitalization of traditional and local knowledge and the decentralization of technological opportunities through participatory and democratic mechanisms. It is notable that the IAASTD report was not fully endorsed by the governments of the USA, Canada or Australia, all of which have highly productivist, intensive agricultural systems. Nevertheless, the IAASTD report offers support to the many organizations and individuals campaigning for sustainable agriculture to be recognized as a fundamental building block of a sustainable world future.

Conclusion

This chapter has provided a broad overview of the main sustainability issues facing agriculture and the ways in which solutions to these are embedded within contradictory ideologies and world views. Few would disagree with Robinson's (2009) view that in general terms, sustainable agriculture would include maintenance of soil and water resources, preserving biological and ecological integrity of systems, ensuring that farming systems are economically viable and that they meet the social and cultural norms of the population. There is also increasing recognition that the three pillars of sustainability – the social, economic and environmental – need to be incorporated into discussions about how to build sustainable agriculture. However, the question of how to operationalize this general vision of sustainable agriculture is complex, because agriculture is implicated in a complex web of political, socio-economic and environmental processes operating at a variety of different scales.

Advocates of industrialized agriculture and increased trade liberalization of agricultural products argue that food has never been more plentiful, affordable, and of such consistent quality. They propose that agriculture – with the aid of biotechnological advances such as genetic modification – will continue to become ever more efficient and competitive, thus driving down food prices, conserving resources and eventually increasing access to food for all. Whilst there are many variations in the suggested alternatives to industrialized, corporate-controlled agriculture, critics argue – broadly speaking – that agriculture would best be organized to meet the needs of people and their communities, rather than the shareholders of agribusinesses and biotechnology companies. It is interesting to note that many of these 'oppositional' or 'progressive' agendas for sustainable agriculture are premised on the same principles as those used in post-conflict recovery, such as the emphasis on small-scale farms, democratic participation in resource management, and the provision of enabling governance and physical infrastructure. Indeed, as illustrated in some of the chapters in this book, it is also likely that the innovative solutions developed from force of necessity in post-conflict situations may offer insights into how sustainable agriculture could be operationalized in other parts of the world.

References

Altieri, M. 2008. Small farms as a planetary ecological asset: five key reasons why we should support the revitalization of small farms in the Global South. Institute for Food and Development Policy [Online]. Available at: http://www.foodfirst.org/en/node/2115 [accessed: 1 November 2011].

Bailey, R. 2011. *Growing a Better Future: Food Justice in a Resource-Constrained World*. Oxford: Oxfam International.

Barrett, H.R., Browne, A.W. and Ilbery, B. 2004. From farm to supermarket: the trade in fresh horticultural produce from sub-Saharan Africa to the United Kingdom, in *Geographies of Commodity Chains*, edited by A. Hughes and S. Reimer. London: Routledge, 19–38.

Beddington, J. 2010. Global food and farming futures. *Philosophical Transactions of the Royal Society, B 2010*, 365, 2767.

Benton, T., Dougill, A., Fraser, E. and Howlett, D. 2011. The scale for managing production vs. the scale required for ecosystem service production. *World Agriculture: Problems and Potential*, 2(1), 14–21.

Blay-Palmer, A. (ed.). 2010. *Imagining Sustainable Food Systems: Theory and Practice*. Farnham: Ashgate.

Blewitt, J. 2008. *Understanding Sustainable Development*. London: Earthscan.

Busch, L. 2010. Can fairy tales come true? The surprising story of neoliberalism and world agriculture. *Sociologia Ruralis*, 50(4), 331–51.

Center for Stabilization and Reconstruction Studies (CSRS). 2010. *Agriculture: Promoting Livelihoods in Conflict-Affected Areas*. Monterey, California: CSRS.

Dixon, J. and Gulliver, A. with Gibbon, D. 2001. *Farming Systems and Poverty*. Rome: FAO and the World Bank.

Dowler, E., Kneafsey, M., Lambie, H., Inman, A. and Collier, R. (in press). Thinking about 'food security': engaging with UK consumers. *Critical Public Health*.

ETC Group. 2009. *Who Will Feed Us? Questions for the Food and Climate Crises, Communiqué 102*. Available at: http://www.etcgroup.org/upload/publication/pdf_file/ETC_Who_Will_Feed_Us.pdf [accessed: 25 September 2011].

Food and Agriculture Organization of the United Nations (FAO). 2003. *Trade Reforms and Food Security: Conceptualizing the Linkages*. Rome: FAO.

Food and Agriculture Organization of the United Nations (FAO). 2010. Global hunger declining, but still unacceptably high. Available at: http://www.fao.org/docrep/012/al390e/al390e00.pdf [accessed: 25 September 2011].

Foresight. 2011. *The Future of Food and Farming*. London: Government Office for Science.

Goodman, D. and Du Puis, M. 2002. Knowing food and growing food: beyond the production–consumption debate in the sociology of agriculture. *Sociologia Ruralis*, 42, 5–22.

Hendrickson, M. and Heffernan, W. 2002. Opening spaces through relocalization: locating potential resistance in the weaknesses of the global food system. *Sociologia Ruralis*, 42, 347–68.

Hinrichs, C. 2010. Conceptualizing and creating sustainable food systems: how interdisciplinarity can help, in *Imagining Sustainable Food Systems: Theory and Practice*, edited by A. Blay-Palmer. Farnham: Ashgate.

Huang, H., von Lampe, M. and van Tongeren, F. (2011) Climate change and trade in agriculture. *Food Policy*, 36, S9–S13.

International Assessment of Agricultural Knowledge, Science and Technology for Development (IAASTD). 2009. *Agriculture at a Crossroads*. Washington, DC: IAASTD.

Jarosz, L. 2011. Defining world hunger: scale and neoliberal ideology in international food security policy discourse. *Food, Culture and Society*, 14(1), 117–36.

Koç, M. 2010. Sustainability: a tool for food system reform?, in *Imagining Sustainable Food Systems: Theory and Practice*, edited by A. Blay-Palmer. Farnham: Ashgate.

La Via Campesina. 2010. *Sustainable Peasant and Family Farm Agriculture Can Feed the World*. [Online]. Available at: http://viacampesina.org/en/index.php?option=com_content&view=article&id=1044:sustainable-peasant-and-family-farm-agriculture-can-feed-the-world.

La Via Campesina. 2011. What is La Via Campesina? [Online]. Available at: http://viacampesina.org/en/index.php?option=com_content&view=category&layout=blog&id=27&Itemid=44 [accessed: 25 September 2011].

McMichael, P. 2011. Food system sustainability: questions of environmental governance in the new world (dis)order. *Global Environmental Change*, doi:10.1016/j.gloenvcha.2011.03.016.

MacMillan, T. and Dowler, E. 2011. Just and sustainable? Examining the rhetoric and potential realities of UK food security. *Journal of Agricultural and Environmental Ethics* [Online]. Available at: http://www.springerlink.com/content/0537k617113tk871.

Monsanto. 2011. Sustainable Agriculture. [Online]. Available at: http://www.monsanto.com/ourcommitments/Pages/sustainable-agriculture.aspx?WT.mc_id=1_susag [accessed: 24 September 2011].

Naylor, R. 2011. Expanding the boundaries of agricultural development. *Food Security*, 3, 233–51.

Popkin, B. 2003. The nutrition transition in the developing world. *Development Policy Review*, 21(5–6), 581–97.

Redclift, M. 2005. Sustainable development (1987–2005): an oxymoron comes of age. *Sustainable Development*, 13(4), 212–27.

Robinson, G. 2009. Towards sustainable agriculture: current debates. *Geography Compass*, 3/5, 1757–73.

Watson, R. 2011. Strategic priorities for commissioning and use of evidence. London: Department for Environment, Food and Rural Affairs (DEFRA).

Weis, T. 2007. *The Global Food Economy: The Battle for the Future of Farming*. London: Zed Books.

World Food Programme. 2011. Hunger stats. [Online]. Available at: www.wfp.org/hunger/stats and http://www.wfp.org/hunger/faqs [accessed: 10 September 2011].

Chapter 2

The Impact of Conflict on Agriculture and Post-conflict Reconstruction Challenges

Alpaslan Özerdem and Rebecca Roberts

Introduction

Violent conflict disrupts every aspect of public and domestic life, causing widespread death and injury and damage to personal property, buildings, infrastructure and natural resources. The insecurity and uncertainty lead to large-scale displacement of civilian populations and psychological trauma, undermining social structures, coping mechanisms, and the acquisition of formal and informal education and skills. Violence, or the threat of violence, prevents civilians from undertaking their daily activities and from travelling to work, school or the marketplace. Systems of governance and rule of law lack the strength and legitimacy to implement and maintain public services, develop national strategies, promote livelihood security and economic growth, and protect internal and international trade.

The interactions between conflict and agriculture, and between agriculture and human life are complex and dynamic. Agricultural production, essential to food security and livelihoods, is affected by conflict directly, through the destruction of natural resources and neglect, and indirectly, through the broader impacts of conflict. Conflict is considered to be the most common cause of food insecurity (Erskin and Nesbett 2009). Although it is not always possible to quantify the impact conflict has on agriculture in human or economic terms, or to separate the direct and indirect impacts of the conflict on agriculture from other social, natural or economic factors, it is possible to explore the various ways in which agriculture is affected in conflict and post-conflict environments. Exactly how it is affected depends on the specific context, but there are common themes.

Drawing on examples from around the world, this section provides a brief overview of the impact conflict has on agriculture. The discussion is organized around the overlapping environmental, social, economic and governance issues disrupted by conflict and related to agriculture. Although these issues are not discrete they illustrate how the context in which agriculture takes place is damaged and disrupted by conflict.

Environmental Issues

Physical damage, either deliberate or as a consequence of the violence, is the most obvious impact of conflict. Produce in the field or in storage may be burned, stored seed for the next planting season destroyed, water resources polluted, equipment damaged and livestock killed. Agriculture is targeted to deny food, weaken morale and physical strength, cause immediate damage to livelihoods and to impede the rehabilitation of the agriculture sector, which hampers post-conflict reconstruction processes in general. Targeting of agriculture has been used as a weapon of war throughout the world. In Sudan, governments in Khartoum deliberately deprived the Dinka of food by denying access to land to weaken their resistance (Messer, Cohen and D'Costa 1998). In Lebanon in 2006, local and national trade was disrupted because aerial bombing meant that farmers were afraid to transport their harvested crops to market and the Israeli-imposed sea blockade prevented produce from being exported.

Large-scale aerial bombing, such as that in South-East Asia during the Vietnam War, causes immediate damage to crops and kills or maims livestock. Chemicals remaining after bombing can contaminate the soil and water, posing a threat to human life. Bombs that fail to explode on impact continue to deny access to natural resources and impede post-war recovery (UNIDIR 2007). Similarly, landmines prevent safe access to land for crop production, grazing, water, wood and other resources long after fighting has ceased. Such contamination can alter behaviour, affecting the livelihoods of people living close to minefields. In Cambodia, agricultural production was abandoned by some living in mined areas in favour of hunter-gatherer activities which traditionally supplemented food production rather than provided the main food source (Davis 1994 cited in Harpviken and Isaksen 2004). In Kosovo, cluster munition contamination forced herders to graze animals on land traditionally reserved for crops, leading to a reduction in agricultural production (ICBL 2007).

Agriculture can be damaged during conflict because in a fight for survival, people engage in unsustainable environmental practices (UNEP 2007). Travel to some fields may be impossible so the usual crop rotation is not implemented. People concentrate on growing fast-yielding crops which drain the soil of nutrients, and are less likely to invest time in growing crops such as citrus or olives which take several years before the first harvest. Migrating populations can destroy both natural resources and the cultivated land of others en route. Refugee camps can cause environmental damage because of the overexploitation of natural resources such as water and wood, which can lead to drought and soil erosion, and create tensions between the refugee and host communities (Black 1994, Zetter 1995).

Infrastructure is deliberately damaged or falls into disrepair during conflict, which directly affects agricultural production in the fields, mechanized farming operations because there is a lack or fuel or spare parts, factories for processing produce, and access to markets. In Afghanistan, for example, most of the population in rural areas depends on irrigated land for its agricultural production, yet the extensive formal and traditional irrigation systems that distribute ground and surface water for crop production were severely damaged during the decades of conflict and are no longer effective (Rout 2008). The lack of water has reduced yield and both the size and the number of the locations that can be farmed.

Social Issues

Conflict causes massive displacement and significant changes in the demographic structure, social practices and aspirations – all of which affect agriculture. Large-scale displacement leads to the neglect of produce, which may die in the field or fail to meet expected yields. If the conflict is short, rehabilitating crops after a period of neglect is relatively simple and the worst impact may be the loss of one harvest. However, the neglect of livestock even for a short period can have a long-term impact because it takes time to replace animals that have been killed, and for animals to regain the condition lost during a period of neglect (UN FAO 2006). Rural populations draw on savings and resources to survive in the short term, making them vulnerable to future shocks. The loss of income from harvest and the loss of equipment and other assets destroyed in conflict creates indebtedness as people have to borrow money to reinvest in their livelihoods after exhausting their own reserves (Messer, Cohen and D'Costa 1998).

Often it is the most able who flee to safety, leaving the elderly and infirm to tend the land. Rural areas may be deserted by men of working age because they leave their families to join the fighting, are forcibly recruited as combatants, or flee before they are conscripted. This results in the women having the double burden of domestic responsibilities and agricultural production. Women may continue to bear the double burden after the war if male family members have been killed. Those injured during conflict may struggle physically to contribute to the household income through agricultural production so they become an economic burden. Following conflict, it is often the men who are the first to return home, or to explore a new area to assess its potential for agriculture. As a result, men of working age are most likely to be killed or injured by landmines or unexploded ordnance. Conflict often leads to high numbers of female-headed households and to women, the young and the old generating an income from agriculture to care for disabled male family members.

Protracted displacement leads to a loss of traditional agricultural knowledge and practice because it is not passed on through the generations. There is often increased urbanization, so children born to parents from rural areas in urban environments have no knowledge and little interest in undertaking agriculture. Displaced populations from southern Sudan who have been living in Khartoum for two decades report that their children reject the agricultural lifestyle of their parents (author field research, Khartoum, Sudan, April 2004). Similarly, in the Balkans, displaced rural populations were reluctant to return to their agricultural income-generating activities after the war, having experienced a more physically comfortable life in the city.

Traditional farming know-how that is specific to particular environments can be lost. At the same time, formal education is disrupted, resulting in a population that is less able to assume responsibility for its own agricultural development, engage in research to improve farming practices, and ensure that its agriculture sector is internationally competitive.

Economic Issues

Damage to the agricultural economy at the national and local level is immediate and devastating and can take years to repair. Protracted civil conflict in Angola meant that a country with the natural resources for a productive agriculture sector saw agricultural production that contributed 29 per cent to gross national product in 1991 fall to just 6 per cent by 2000 (EIU 2002 cited in ICG 2003). At a local level, the lack of plentiful or profitable agricultural work during or immediately after conflict forces men of working age in particular to look elsewhere for employment. If successful, it is unlikely that those who previously worked as agricultural labourers will take up the occupation again. The gap in the labour market leaves a vacuum, to be filled by the less physically able, or by foreign workers who send their wages back to their home country rather than investing in the local economy. The loss of the traditional labour force changes the demographic structure and is likely to result, in the short to medium term at least, in agricultural production that is less profitable to the local population.

The destruction of roads and bridges prevents access to local, national and international markets for small- and large-scale agricultural producers. In Angola, the infrastructure is so poor there is little point in households growing more than they can consume themselves or sell to their immediate neighbours (author field research, Moxico Province, Angola, December 2010). Landlocked countries such as Afghanistan and Nepal have particular difficulty in exporting goods because they depend on good relations with their neighbours, and the willingness of international markets to pay a premium for produce which is costly to export. Conflict which destroys the internal infrastructure makes this international trade even more difficult. The lack of access to markets can lead to shrinkage of the agriculture sector because there is no incentive to engage in agricultural production beyond the subsistence level.

The value of agricultural produce for export can be increased significantly if there is national capacity to process the produce in some way. Countries that export unprocessed produce such as Cambodia are in fact selling the profits on to a second country. During conflict, the time and resources to add value through processing do not exist. There is an immediate need to generate an income, processing plants may be damaged or destroyed, and the spare parts and energy needed to maintain them may be unavailable. Following a conflict, countries lack the resources to invest in processing.

Governance Issues

Traditional and Western-style governance structures and practices are either the target of conflict or are damaged during conflict. To have a coherent national development plan for sustainable agriculture that can be implemented effectively there must be stability, and a respected and legitimate authority. During conflict, the environment is unstable and authorities are too weak to initiate agricultural programmes, which, given their long-term nature, are unlikely anyway to be a priority. National schemes to improve agricultural production, through education or provision of resources, grind to a halt. As a result, agricultural development can be put back decades, and the level of investment in time and resources needed to rehabilitate agricultural production to match pre-conflict levels can be huge. Calculations based on data collected from sub-Saharan African countries in conflict from 1970 to 1990 suggest that agricultural production is reduced on average by 12 per cent during each year of the conflict. In Angola, conflict reduced agricultural production by as much as 44 per cent (Messer, Cohen and D'Costa 1998: 19–21). Countries, similar to Nepal, that were self-sufficient before conflict become reliant on food imports during and after conflict because of an inability to increase food production to meet the population's needs.

Official peacetime government initiatives to provide practical support to farmers are often suspended. For example, before the conflict in Nepal the government supplied farmers with fertilizers but restrictions on mobility by the Maoists put a stop to this (Shakya 2009). Investments made in agricultural extension work were also lost as agricultural research centres were deliberately targeted and data and equipment destroyed. As a consequence, skilled staff members left.

Conflicts can undermine traditional governance systems for managing shared natural resources (Castro and Nielsen 2003: 2). In parts of Sudan, the traditional practices of pastoralists and farmers have enabled both groups to coordinate their use of the same piece of land. The system is not perfect but traditional conflict-management measures have allowed pastoralists and farmers to negotiate access to the land for their livelihood activities. The decades of civil war in Sudan have disrupted these traditional livelihood patterns and conflict-management mechanisms, exacerbating tensions between the two groups. The situation has been made worse by rapid desertification which has reduced the amount of land suitable for agriculture and livestock (African Rights 1995).

Modern governance systems which operate within national boundaries also clash with traditional pastoralist practices. National boundaries often cut across pastoralist lands, dividing the same community among several countries. Political borders affect access to resources and can have a devastating impact on pastoralist groups. At the same time, the movement of the communities and their livestock creates border tensions between countries (Nori, Taylor and Sensi 2008, Toulmin, Hesse and Cotula 2003).

Governance systems regulate land rights. The lack of good and consistent governance results in disputes over land and dispossessed populations. In Sudan's Nuba Mountains, decades of conflict have created multiple systems of land rights. Colonial powers redistributed land ignoring deep-rooted customary laws. Despite the decades spent in implementing another system, customary law continues to regulate some land access. In the 1970s and 1980s, in a bid to win and guarantee the loyalty of allies, governments in Khartoum gave away large tracts of land in the Nuba Mountains for commercial farming, dispossessing the local population. However, the civil war led to the abandonment of these grandiose farming schemes. By the time of the Comprehensive Peace Agreement in Sudan in 2005, in the Nuba Mountains, ownership of land and the system used to regulate that land is unclear (author field research, Nuba Mountains, Sudan, 2004 and 2006).

Cambodia is another example where conflict has dispossessed rural communities of their land. The Khmer Rouge appropriated all land, declaring that there could be no private ownership. After the conflict, with massive displacement and large numbers of deaths, the original landowners may no longer be alive or in the country. Corruption in the post-war period has led to land grabbing by the elite for farming, building or establishing commercial enterprises such as casinos and golf courses on the Thai border. Untangling who owns what after protracted periods of conflict, when multiple previous systems of landownership have been used but have not been universally accepted, is a long-term challenge (author field research, Cambodia, 2003 and 2005).

External assistance during conflict can have a long-term negative impact on agriculture. For example, the provision of food creates a dependency culture and there becomes an expectation that the international community will meet basic needs such as food. Consequently, the impetus to be self-sufficient is removed, and the skills necessary to farm effectively are lost as well as the ability to be self-reliant. In Nepal, the provision of rice by the World Food Programme over a number of years has changed the staple diet of the population. Before the conflict people grew other crops and relied on them as their main calorific source. Rice was something that was reserved for special occasions and eaten only rarely. By 2010, the population had come to expect to eat rice and traditional staple foods were seen as inferior (author field research, Kathmandu, Nepal, 2010). Most areas of Nepal are unsuitable for rice production and those where rice can be grown cannot provide enough to feed the whole country. Consequently, the country relies on importing rice, which is expensive and means that the market for locally grown produce is reduced. It is suggested that communities in some parts of the country go through an annual ritual of planting rice with the expectation that the harvest will fail. Once the crop has failed, food aid is delivered.

Overall, it is evident that conflict impacts on the agricultural sector, which has a broader impact on many aspects of public and private life, and thus communities emerging from armed conflicts face a wide range of challenges. Unless they are addressed effectively, food security and economic recovery in many post-conflict contexts will be impossible to achieve. The following section focuses on the concept of reconstruction and characteristics of challenging environments faced by war-torn societies.

Post-conflict Reconstruction Challenges

As a result of the devastation typically wrought by an armed conflict, post-conflict environment conditions are rarely conducive to begin a complex process of rebuilding a new society. All societies differ, and the same goes for post-conflict environments in terms of their environmental, social, economic and governance characteristics. However, there are features that tend to characterize conflict-affected environments. These factors need to be addressed by initiatives aimed at creating a sustainable peacebuilding process. The common features of war-torn societies can be seen in Table 2.1, which has been adapted from Nicole Ball (1996).

With the complexity of such challenges identified by Ball, an approach of 'rebuilding' what existed before the conflict alone would not be enough. This would mean simply reinforcing structural inequalities and discrimination that created ripe conditions for the conflict in the first place. To investigate this issue further, Table 2.2 lists the reasons identified by Junne and Verkoren (2005) for why a 'rebuilding'-alone approach would not be effective to respond to the needs of post-conflict environments, which underlines the significance of a 'rebuilding plus reform' approach.

Therefore, with such post-conflict challenges in mind, the term 'reconstruction' is often used to indicate three major realms of activities as identified by Barakat (2005):

- Physical/socio-economic and political: Rebuilding infrastructure and essential government functions
- Capacity-building and institutional strengthening: Improving the efficiency and effectiveness of existing institutions
- Structural: Reforming the political, economic, social and security sectors

The key issue to consider here is the need for a holistic approach in understanding post-conflict reconstruction. Rather than tackling its tangible and intangible aspects separately, these three realms would need to be integrated with each other in a single approach. For example, with the rebuilding of the education system it might be essential to rebuild schools, train teachers and reform the national curriculum, but the most effective approach would be to consider all these activities in relation to each other. Similarly, for the rehabilitation of the agricultural sector, it should not be only the provision of material goods but often transfer of know-how and reform of governance structures for better access to natural resources.

Table 2.1   Characteristics of war-torn societies

Environmental characteristics	Social characteristics	Economic characteristics	Governance characteristics	Security characteristics
Environmental degradation	Weakened social fabric	Significant contraction of legal economy and expansion of illegal economy	Weak political and administrative institutions	Bloated security forces with a political role
Damaged infrastructure creating problems of access and communication affecting abilities to trade locally, nationally and internationally	Destruction or exile of human resources and traditional know-how	Reversion to subsistence activities	Vigorous competition for power at expense of attention to governing and security	Armed opposition, paramilitary forces and over-abundance of small arms
Destruction of marketplaces and processing plants	A population traumatized by war	High levels of indebtedness	Limited legitimacy of political leaders and non-participatory political system	Lack of transparency in security affairs and accountability to civil authorities and population
		Lack of clear national economic policies and financial regulation	Lack of consensus on direction country should follow	Need to reassess security environment and structure security forces accordingly
		Unsustainably high defence budgets	Conflicts over ownership of and access to land	History of human rights abuses perpetrated by security forces

Source: based on Ball 1996.

Table 2.2   Why 'rebuilding' is not good enough

	Bias in pre-conflict structures	Desirable difference in post-conflict structures
Security forces	The security forces were often highly biased and more an instrument of suppression than protection for large parts of the population.	Access to employment in the security forces needs to be open to all sectors of the population. Security forces need to see themselves as servants of the people and treat people fairly and equally.
State structures	Government positions were often in the hands of a specific group of the population, and government expenditures were regionally concentrated.	Access to government positions has to be open to all groups; regional infrastructure should be more balanced.
Justice	Justice system was often underdeveloped, badly equipped, biased and corrupt. In a climate of violence, judges are intimidated.	To establish the rule of law, independent judges have to be installed and maintained.
Infrastructure	Infrastructure is often more geared to the exploitation of natural resources than to service; people's denial of access to others.	Infrastructure should provide roughly equal service to people in different regions, and facilitate development of a modern economy that provides larger parts of the population with a chance to earn their living.
Education	Education used to be elite-oriented, with a concentration of universities and schools around capitals and a neglect of most of the country.	Educational facilities need to be boosted, and made equally available across the different groups in society.
Health	Many health systems favour urban-based curative care.	Future health systems provide more equal care for all of the population, and have a strong emphasis on preventive care.

Source: Junne and Verkoren 2005.

In other words, when thinking about post-conflict reconstruction, it is important to bear in mind that reconstruction is fundamentally a developmental challenge. There is often a debate over the distinction between relief and reconstruction that tends to focus on the timing of those activities. Experience in conflict-affected environments shows, however, that reconstruction does not necessarily follow relief. They need, rather, to be undertaken simultaneously. Moreover, there are a host of challenges connected to who sets the agenda of the priorities and process for reconstruction, and what it entails. Based on the post-settlement peacebuilding framework identified by Ramsbotham, Woodhouse and Miall (2011), Table 2.3 presents four critical goals of reconstruction in terms of short-, medium- and long-term measures.

Table 2.3   Post-conflict reconstruction framework

	Short-term measures	**Medium-term measures**	**Long-term measures**
Security sector reform	Disarmament, demobilization of factions, separation of army and police	Consolidation of new national army, integration of national police, and demining	Demilitarization of politics, transformation of cultures of violence
Governance	Establishment of transitional government, constitutional reform	Overcoming the challenge of the second election and establishment of institutional structure of governance in all spheres of life	Establishment of a tradition of good governance including respect for human rights, democracy, rule of law
Socio-economic recovery	Provision of humanitarian relief, essential services of education, health and welfare and communications	Rebuilding of infrastructure, housing and services, livelihoods and employment opportunities, and reintegration of displaced populations	Stable long-term macroeconomic policies and economic management, locally sustainable community development, and distributional justice
Justice and reconciliation	Overcoming initial distrust	Managing conflicting priorities of peace and justice, and recovery of truth	Healing psychological wounds, prosecution of war criminals, long-term reconciliation

Source: Ramsbotham, Woodhouse and Miall 2011.

Although, this reconstruction framework provides an apparently clear-cut list of short-term, medium-term and long-term measures in each of the four categories, the intention is not to portray reconstruction as linear and sequential. Post-conflict environments, as explained above, are complex, and are unlikely to fit neatly into such a simplistic theoretical framework. Political agendas may dictate that some of the medium-term measures are started as interim measures, or vice versa, while some short-term measures might not be tackled until years after the conflict has ended. For example, the disarmament of warring sides is given as a short-term measure in Table 2.3 – which is and needs to be the case in many post-conflict environments. Disarmament is, however, a sensitive undertaking. To persuade warring sides to support the peacebuilding process, it may be necessary for demobilization rather than disarmament to be undertaken in the short term. Disarmament itself may need to wait until a much later stage, once warring sides have developed confidence in the continuation of the peacebuilding process. In the agricultural sector, land reform or clearance of landmines would need to be tackled as short-term measures because this would be the only way that war-affected communities could initiate their agricultural activities.

There are a number of challenges and dilemmas encountered frequently in the post-conflict reconstruction process. First, articulating a vision for the future that is shared by all or most stakeholders can be crucial to the success of post-conflict reconstruction, as it can mobilize populations to contribute to the process effectively. The task of reconstruction requires the development of collaborative structures of governance with the participation of actors from national and local authorities, local NGOs and grassroots-level organizations, the international aid community and the private sector. The absence of any of these actors – particularly local ones – can result in programme failure, long delays in responding to urgent needs, the waste of scarce resources, and, most dangerously, renewed violence. In the contemporary practice of establishing a vision for reconstruction, external actors tend to play a central role, leaving too little space for the inclusion of the views of internal actors about what they wish to see the reconstruction process doing, and how it should take place. Setting up an agriculture development policy in a post-conflict environment would be likely to be dictated by the interests of economic and political elites. The resistance to land reform as part of post-conflict reconstruction is often a key challenge faced by peasant communities in war-affected countries. The way that its political elites are closely linked to multinational companies, for example, also plays a significant role in the implementation of liberal economy measures in its agricultural sector. Such approaches tend to negate local communities' means of recovery and competition, and furthermore create a culture of dependency on external assistance.

Second, reconstruction done well involves far more than meeting a series of set objectives. There are critical issues around not only what reconstruction delivers, but also how it delivers, and when. The process dynamics are arguably among the most important aspects of reconstruction in conflict contexts, yet they are too often neglected. Reconstruction activities are often structured as goal-based rather than process-based. Also, reconstruction programmes are likely to be much more beneficial if they focus on community-owned assets rather than on individually owned assets. For example, the reconstruction of irrigation systems can play a significant role in reasserting inter-group relations and interdependence, whereas the reconstruction of individual family housing may cause conflict-affected populations to compete and compare with each other. The reconstruction process can also be a means for healing societal wounds caused by armed conflict. This can be achieved in a variety of ways. For example, the process of reconstructing key lifelines – such as water supply systems, clearance of landmines from agricultural land or community-based livelihoods programmes – can become opportunities for those conflicting sides to work towards an improved infrastructure, service or means of income that benefits all.

Third, the practice of enabling a participatory environment is often viewed and carried out from an 'international community' perspective, yet post-conflict reconstruction strategies need to include representatives from many different stakeholder groups in open and transparent decision-making. This is vital to ensure an inclusive strategy, and also crucial if the local community is to cooperate with international actors. In establishing such an inclusive approach, reconstruction agencies need to build on the economic interdependence of different ethnic and religious groups, and agriculture can be an enabling context for such community-cohesion-centred objectives. This can help to reduce tension between and within those groups, and create an environment more conducive to participation. However, a top-down approach oriented to physical reconstruction can undermine the dignity of people who have already lost so much in armed conflict, if they are not fully included in the process. Yet the very survival and coping mechanisms of conflict-affected people – for example, the way they deal with food security challenges and continue with their agricultural practices even during the conflict – suggest just how much they can offer to the reconstruction process.

Fourth, it is critical that the reconstruction process yields peace dividends as early as possible. Without such tangible outcomes, it would be difficult to ensure conflict-affected societies' continued support for peacebuilding. Furthermore, there would be increased risk of former combatants deciding to take up their weapons once again – to resume their fighting and/or to secure their livelihood. Once again, agriculture as a key aspect of the economy in many conflict-affected environments could play a pivotal role in providing peace dividends in terms of assisting the rehabilitation of the agriculture sector through the provision of seeds, equipment, fertilizers and other materials to both former combatants and communities in general. This is particularly the case when dealing with the legacy of war economies, as sometimes such environments allow some communities to make a living out of the production of crops like poppy seeds or marijuana, or from smuggling a wide range of goods, from oil and weapons to domestic goods. If the end of conflict removes such sources of revenue without replacing them with peace dividends through new means of employment and livelihoods, then it would prove extremely difficult to convince such communities to support peace, while they wait for better days to come.

Fifth, similarly to the challenge of dealing with war economies, one of the greatest challenges of reconstruction is to reverse economic trends and social attitudes that may have developed during the years of conflict. Armed conflicts distort economies, patterns of production, trade and employment systems and relationships. With the arrival of the international community those patterns are further diverted from their original structures and bases. With time, external relief runs the risk of being misinterpreted as a substitute for local authority and government budget allocations. Moreover, the distribution of such relief assistance as food over a long period often results in a number of negative impacts on the local economy, such as taking all incentives from local farmers to produce so that they abandon their agricultural land and migrate into overpopulated urban areas. Therefore, it is important that even in the provision of relief assistance there should be a long-term vision of reconstruction and the ways of linking one to the other. For example, rather than providing protracted food aid, it would likely be more effective if it was provided as an exchange of labour input by conflict-affected populations in the rehabilitation of agricultural facilities such as irrigation systems. Rather than food aid alone, the provision of seeds and development of nurseries for agriculture as early as possible would be another example of linking relief assistance with long-term reconstruction objectives.

Finally, mobilizing adequate resources is another key issue for the sustainability of reconstruction strategies. Post-conflict reconstruction experiences around the world indicate two major problems. The first relates to the type of resources that are given priority: financial and physical resources are often seen as more important than human and organizational ones. The second challenge relates to the timing and amount of resources provided in the progression from relief to reconstruction. Current practice often focuses on financial resources provided by external actors to internal actors. However, such financial resources for reconstruction generally come with a package of conditions that may range from economic and social requirements to political ones. Consequently, the agendas, aspirations and values of the international community come to dominate the local context. The prioritization of financial resources creates a multi-layered hierarchical system of decision-making in which the quantity of funds to be provided dominates relationships between different agencies. Instead of such a finance-centric focus on mobilizing resources, priority should be given to empowering and enabling local human resources. Such a perspective would place local agencies in the driving seat of the process, supported by external actors when they are needed. If international agencies were to adopt this approach, for instance in the rehabilitation of agriculture sectors, they would need to do far more than simply use local actors for delivery purposes. They would need to ascertain the most empowering means of transferring know-how, experience and financial resources to local actors.

Conclusion

For most conflict-affected communities around the world, the reconstruction of the agricultural sector is probably one of the most significant starting points as this would give them, particularly those in rural areas, a meaningful opportunity of rebuilding their own lives and livelihoods. However, the agricultural sector's fragility, due to its proneness to external factors from climate conditions to the availability of fertile land and other resources, means that its reconstruction is likely to come up against a wide range of post-conflict challenges. Some of these would be due to the conflict's direct impact on agriculture, as summarized in the first part of this chapter. However, other contextual challenges in the realms of politics, socio-economics, governance and security would all be significant determinants for the reconstruction process. Therefore, the impact of the conflict and overall surrounding context must all be considered carefully in response to the reconstruction needs of the agricultural sector.

References

African Rights. 1995. *Facing Genocide: The Nuba of Sudan*. London: African Rights.
Ball, N. 1996. The challenge of rebuilding war-torn societies, in *Managing Global Chaos: Sources of and Responses to International Conflict*, edited by C. Crocker and F. Hampson. Washington, DC: USIP.
Barakat, S. (ed.). 2005. *After the Conflict: Reconstructions and Redevelopment in the Aftermath of War*. London: I.B.Tauris.
Black, R. 1994. Livelihoods under stress: a case study of refugee vulnerability in Greece. *Journal of Refugee Studies*, 7, 4.
Castro, A.P. and Nielsen, E. 2003. *Natural Resource Conflict Management Case Studies: An Analysis of Power, Participation and Protected Areas*. Rome: United Nations Food and Agriculture Organization (UN FAO).
Davis, P. 1994. *War of the Mines: Cambodia, Landmines and the Impoverishment of a Nation*. London: Pluto.
Economist Intelligence Unit (EIU). 2002. *Angola Country Report*. London: EIU.
Erskin, W. and Nesbett, H. 2009. How can agriculture research make a difference in countries emerging from conflict? *Experimental Agriculture*, 45, 313–21.
Harpviken, K.B. and Isaksen, J. 2004. *Reclaiming the Fields of War: Mainstreaming Mine Action in Development*. Oslo: International Peace Research Institute, Oslo (PRIO)/United Nations Development Programme (UNDP).
International Campaign to Ban Landmines (ICBL). 2007. *Landmine Monitor Report: Towards a Mine-Free World*. New York: Human Rights Watch.
International Crisis Group (ICG). 2003. *Angola's Choice: Reform or Regress*, Africa Report no. 6. Luanda/Brussels: ICG.

Junne, G. and Verkoren, W. (eds). 2005. Postconflict development: meeting new challenges. Boulder, CO: Lynne Rienner Publishers.

Messer, E., Cohen, M.J. and D'Costa, J. 1998. *Food from Peace: Breaking the Links between Conflict and Hunger*. Washington, DC: International Food Policy Research Unit (IFPRU).

Nori, M., Taylor, M. and Sensi, A. 2008. *Browsing on Fences: Pastoral Land Rights, Livelihoods and Adaptation to Climate Change*. London: International Institute for Environment and Development.

Ramsbotham, O., Woodhouse, T. and Miall, H. 2011. *Contemporary Conflict Resolution*. 3rd Edition. London: Pluto Press.

Rout, B. 2008. *How the Water Flows: A Typology of Irrigated Systems in Afghanistan*. Kabul: Afghanistan Research and Evaluation Unit (AREU).

Shakya, A. 2009. *Social Impact of Armed Conflict in Nepal: Cause and Impact*. SIRF.

Toulmin, C., Hesse, C. and Cotula, L. 2004. Pastoral common sense: lessons from recent developments in policy, law and practice for the management of grazing lands. *Forest, Trees and Livelihoods*, 14, 243–62.

United Nations Environmental Programme (UNEP). 2007. *Lebanon: Post-Conflict Environmental Assessment*. Nairobi: UNEP.

United Nations Food and Agriculture Organization (UN FAO). 2006. *Lebanon: Damage and Early Recovery Needs Assessment of Agriculture, Fisheries and Forestry*. Rome: UN FAO.

United Nations Institute for Disarmament Research (UNIDIR). 2007. *The Humanitarian Impact of Cluster Munitions*. Geneva: UNIDIR.

Zetter, R. 1995. *Shelter Provision and Settlement Policies for Refugees: A State of the Art Review*. Uppsala: Nordiska Afrikainstitutet.

Chapter 3
Projections on Future Challenges: Climate Change

Phil Harris and Alpaslan Özerdem

Introduction

As a general term, climate change refers to fluctuations in climatic parameters of a system over a long or medium term of at least decades. The term now is frequently used to refer to that climate change that has taken place since the early twentieth century almost certainly as a result of human activity (Masters, Baker and Flood 2009). The primary underlying cause of this change is the emission of greenhouse gases such as carbon dioxide and methane, which in turn leads to increased globally averaged temperatures (IPCC 2007). It is not the emission of greenhouse gases per se, but their impact on global warming that is most concerning. Although the term climate change is often considered synonymous with the term global warming, climate change encompasses changes in many parameters secondary to, or consequent upon, global warming.

In addition to a general increase in global temperature, climate change significantly increases climatic variability, especially of temperature, precipitation and tropical cyclone activity. Increased temperature variation includes fewer cold days and nights, more frequent hot days and nights, and increased frequency of heatwaves (IPCC 2007). Melting of the ice caps and retreat of glaciers will lead to sea level rise. As a result of changes in the seasonal pattern, intensity and volume of rainfall, the frequency of droughts and of extreme precipitation events leading to floods is increased (Masters, Baker and Flood 2009).

Such a global change in climate is likely to have major impacts on war-torn societies in terms of resource management, dynamics of migration and displacement, adaptation of farming techniques and methods according to changes in weather patterns, protection of social cohesion among rural communities and resilience against possible rapid physical and socio-political changes. Therefore, this chapter will explore what the climate change dynamics would mean for an agricultural sector emerging from an armed conflict and what could be done to prepare war-torn societies as part of post-conflict reconstruction strategies.

Implications for Agricultural Communities

Climate change has profound implications for world agriculture and hence food supply. The climate of a location is perhaps the most important factor determining agricultural land use, including which crops are grown and which livestock are produced, the seasonal agricultural cycle, rotations and the optimum practices for crop production and protection. The effects of climate change are already being experienced by farmers in many regions. For example, modest increases in global temperatures are likely to increase overall crop yields. However, higher temperatures projected to occur within decades with fewer cold days and nights, and more frequent hot days and nights, may increase yields in colder environments but will certainly decrease yields in warmer environments (IPCC 2007). Elevated temperatures and increased frequency of heatwaves will also increase the danger of fires to agriculture. In general, warmer temperatures and less extreme cold periods will tend to increase the incidence and severity of pest damage to crops and livestock.

Increased precipitation in some areas may increase crop yield. On the other hand, excessive or very heavy rainfall can directly damage crops, erode soils, remove essential nutrients and lead to waterlogging which damages soils, directly reduces crop growth and limits the opportunities for mechanical cultivation and harvest (IPCC 2007, Masters, Baker and Flood 2009). Elevated water tables in coastal areas resulting from sea-level rise can result in sea-water inundation of soils and salinization of ground and irrigation waters, with adverse effects on crops. In areas with reduced precipitation, drought will lower crop yields, increase livestock death and contribute to the risk of fires. Increased incidence of tropical storms will also have a direct negative impact by damaging crops. In other words, while agriculture in some areas will benefit from the projected changes in climate, there remains little doubt that many regions, especially in the tropics, will suffer significantly decreased agricultural productivity and increased and more frequent food shortages. IFPRI (2010) predicts that the area of land suitable for crop production will fall by 0.3 per cent and world food production will fall by 2.7 per cent by 2050 under a scenario of moderate climate change.

The effects of climate change on both total agricultural production and short-term variation in yield, with more frequent crop failures, will likely be reflected in upward pressure on and increased volatility of world food commodity prices (Masters, Baker and Flood 2009). There is a real threat that climate change will lead to increasing food insecurity, poorer diets and increased malnutrition, while the additional direct effects of increased temperatures and reduced access to potable water will combine to increase a range of disorders and diseases damaging to health (Smith and Vivekananda 2009). Greater frequency of natural disasters will lead to increased risk of damage to property and life. Consequently, a reduction in agricultural livelihoods in the areas most affected by climate change will certainly lead to increased economic migration from rural to urban areas, and from more to less affected areas in search of more productive lands for crops and livestock. Where the adverse effects of climate reach critical levels that make local livelihoods unviable, wholesale migration of communities with the multitude of problems that entails is very likely (Smith and Vivekananda 2009).

Climate Change as a Source of Conflict

The undoubted impact that climate change can have on the availability and distribution of essential resources including productive land, water and food has a serious potential to initiate or exacerbate conflict between individuals, groups and states (Bruck and Schindler 2009). This is particularly so against a background of poverty, weak governance, political marginalization and corruption in many of the countries most threatened by climate change (Smith and Vivekananda 2009). Competition for control of good-quality water supplies for irrigation, livestock watering and human consumption is already a significant source of conflict between communities and states (Masters, Baker and Flood 2009). Smith and Vivekananda (2009) consider that migration encouraged by the adverse effects of climate change, as with that for other reasons, can have beneficial effects for the migrants and destination communities but it can also be a source of potential conflict by transferring the risk to other areas such as large cities already themselves threatened by sea level rise, or rural areas already only marginally viable. Overall, climate change and its impact on the agriculture sector could be seen as a potential catalyst that might trigger future conflicts over shared natural resources such as water and land.

Such an environment of competition over natural resources would be particularly intensive in post-conflict environments, as the peacebuilding process itself could be considered as the restructuring of 'competition' in politics, economics and other aspects of governance. Ramsbotham, Woodhouse and Miall (2011), in fact, describe peacebuilding as 'Clausewitz in reverse' to point out that the politics of armed conflict continue well into post-conflict environments, but just fought through different means. In other words, one of the key characteristics of post-conflict environments is the way that they experience a high level of competition because of shortages in resources and capacities due to the damage incurred by armed conflict, weak governance structures, and societal divisions and political polarization that are created or exacerbated by the conflict. Therefore, such post-conflict environments are particular vulnerable to new conflict dynamics such as the impact of the climate change on the agricultural sector.

Not being able to deal with possible emerging conflicts over the common use of natural resources would mean that post-conflict societies would be caught unprepared to cope with new conflicting dynamics in their socio-political and economic structures. Even those conflicts that could otherwise be managed effectively through formal and traditional conflict management and resolution mechanisms could become significant challenges for the sustainability of peace. Societal divisions exacerbated by the conflict could easily turn disagreements over natural resources into ethno-religious armed struggles. Therefore, climate change as a source of conflict is particularly critical for societies emerging from civil wars and it is imperative that the medium to long term planning in the agriculture sector in such areas should consider how climate change would affect that particular environment and how the sector could be prepared for such a slow-onset disaster.

Vulnerability to Climate Change

DFID (2005) estimated that 2.5 billion people in developing countries depend on agriculture for their livelihoods. Geographically, the worst effects of climate change will be experienced in the drier areas of Africa and parts of South Asia, where very large numbers of people rely on rain-fed agriculture and where yields of some staple crops such as maize, rice and millet are predicted by IFPRI (2009) to decline by 5–22 per cent and by others by as much as 30 per cent by 2030 (Lobell et al. 2008). In general, less developed countries have a lower capacity to adapt to climate change than developed countries and are thus more vulnerable (Thomas and Twyman 2005).

Within countries some regions are likely to be more affected than others, with delta and other coastal areas, and arid and semi-arid areas particularly affected. The latter areas are predicted to suffer the greatest increase in temperature and moisture stress and, consequently, the greatest reduction in productive land area and crop yields. More sophisticated estimates of hot spots of vulnerability to climate change in sub-Saharan Africa using a range of indicators based on the livelihoods approach highlighted particularly high vulnerability in many regions including mixed arid/semi-arid systems in the Sahel, arid/semi-arid rangelands systems in parts of East Africa, the systems of the Great Lakes region of East Africa, the coastal regions of East Africa and the drier zones of southern Africa (Thornton et al. 2008). Elsewhere, especially in parts of Asia, rising sea levels and flooding represent the main challenge. In other words, it is important to note that most environments that are currently affected by armed conflicts are also those regions of the world that are likely to face new pressures imposed by the climate change.

The adverse effects of climate change are likely to be felt most by the poorest and most vulnerable members of society (Smith and Vivekananda 2009) as they often depend on agriculture for subsistence and almost invariably have limited resources. Low income and limited resources make individuals and communities less able to cope with short-term fluctuations in agricultural production or food prices by purchasing food, and less able to invest in climate change mitigation technologies. The agricultural systems practised by poor farmers often rely on traditional management of complex natural or semi-natural ecosystems for livestock production or have a greater dependence on stable weather patterns than industrialized agriculture, potentially making them more vulnerable to climate change (Thornton et al. 2008).

Several authors have considered the relationship between climate change and vulnerability of resource-poor farmers and reported on their resilience to such change (O'Brien et al. 2004). Thornton et al. (2008) describe two approaches to vulnerability, summarized as the 'end point' and 'starting point' approaches. In the 'end point' approach, vulnerability is determined by climate change impact mediated by capacity to adapt. The 'starting point' vulnerability has been viewed by Thornton et al. (2008: 27) as 'a state that is governed not just by climate change but by multiple processes and stressors'. These authors consider two types of vulnerability: biophysical vulnerability, which concerns the sensitivity of the natural environment, and social vulnerability, the sensitivity of the human environment to climate change.

Adapting to Climate Change

Conditions that lead to effective response to climate change are listed by IPCC (2007) as economic resources, infrastructure technology, infrastructure, information and skills, institutions and equity, and subsequently also human capital and governance structures, and by IFPRI (2010) as including climate change awareness, access to rural services such as credit and extension, secure land rights, agricultural technology, and communication. In a similar vein, the Africa Climate Change Resilience Alliance (ACCRA) framework reviewed by Jones et al. (2010) considers the adaptive capacity at the community level to rely on the asset base, institutions and entitlements, knowledge and information, innovation, and flexible forward-looking governance.

Adaptations by individual farmers include modification of agricultural practices and adoption of new crops, cultivars and livestock species and breeds. General priorities are listed by DFID (2005) as 'increased knowledge and application of new irrigation technologies, water conservation measures, drought resistant varieties, new cropping patterns, improved management of pests and diseases through integrated pest management, crop diversification particularly with alternative crops'. Many of the required technologies, including drought-tolerant cultivars, drip irrigation and minimum tillage (DFID 2005), intercropping, increasing biodiversity, utilizing surrounding ecosystems and their services, integrated crop and pest management, and targeted fertilizer application (Masters, Baker and Flood 2009), have been extensively researched over many decades and are already being more or less widely implemented (DFID 2005). The constraints to adoption are often poor extension and promotion rather than technical limitations per se.

In an analysis of studies on the perception of, and adaptation to, climate change by farm households in sub-Saharan Africa carried out by the International Food Policy Research Institute, Ringler (2010) clearly demonstrates that the impacts of climate change are occurring and apparent to farmers in sub-Saharan Africa from Ethiopia to Kenya and South Africa. Ringler (2010) also pointed out that the perceptions of current climate change differed between farmers in different circumstances. In an area of South Africa, while a vast majority of farmers reported increased temperatures and decreased rainfall, two-thirds had done nothing to adapt while the other third irrigated more, used water-harvesting systems, planted different crops, changed planting dates and provided supplemental livestock feed. In contrast, in the Nile Basin area of Ethiopia, while a smaller proportion, though still a majority, of farmers reported temperature and rainfall change, more had changed their farming practices to include soil conservation efforts, new crop varieties and tree planting. In semi-arid Kenya more than 95 per cent of interviewed farmers reported increased temperatures and increased rainfall variability and 88 per cent had taken steps to adapt by growing different crop varieties, changing crops or planting dates and adopting agroforestry practices (Ringler 2010).

As well as highlighting the social and economic environments prerequisite for adaptation to climate change in agricultural communities in less developed countries, analyses have also highlighted the main constraints faced by households. Ringler (2010) concludes that the ability of farmers to adapt is compromised by lack of access to credit, markets, information and other resources. According to IFPRI data, the poorest households are the least likely to adapt because they do not own livestock, rely on rain-fed agriculture, have limited land, poor soils and inadequate information, and lack property rights, credit, labour, market access, water and other inputs (Ringler 2010). Many of the regions of the world most threatened by climate change occur in fragile states with poor infrastructure, ineffective state institutions, little social and economic support for individuals and communities adversely affected by climate change, and a high risk of political instability and violent conflict. It is also important to stress the need to focus on the social, economic and political consequences of climate change, and that social, economic and political issues influence the impact of climate change on people's lives (Smith and Vivekananda 2009).

Post-conflict Situations and Climate Change

In post-conflict communities, climate change adds further problems and interacts with other problems to worsen an already difficult situation. Whatever approach is used to consider vulnerability of households to climate change, those in post-conflict situations are likely to have greater vulnerability because of lower adaptive capacity (end point approach) and/or greater social vulnerability (starting point approach), as defined by Thornton et al. (2008), above. In post-conflict situations many of the conditions that support farmer adaptation to climate change are particularly disrupted. In many traditional societies, the structures of extended family and community are the backbone of coping mechanisms against adversaries such as disasters and other external impacts. However, heavy death toll and displacement would mean that in post-conflict environments it would be more difficult for communities to rely on such kinship and traditional structures.

There have been attempts to develop indicator-based assessment of climate change vulnerability and to produce a vulnerability index. Gbetibouo and Ringler (2009) assessed the vulnerability of South Africa's nine provinces to climate change using a set of indicators related to exposure to hazards, sensitivity to hazards and adaptive capacity. Such an approach would be useful in the post-conflict situation to indicate the impact of conflict-induced change on climate change vulnerability and to highlight areas for priority intervention to improve climate change adaptability. A vulnerability measurement tool would be particularly useful in the preparedness of the agricultural sector as part of post-war recovery programmes. However, in the preparation of such an assessment, the lack of baseline data in areas emerging from protracted armed conflicts needs to be considered in making decisions on the methods and techniques of measurement.

Many agricultural communities in less developed countries depend on semi-natural and natural ecosystems for livestock production, wild-harvested food collection, non-timber forest products and fuel gathering. These agroecological relationships rely heavily on indigenous knowledge. These ecosystem services are liable to disruption both by climate change and by conflict. Many of these relations also depend on informal social and economic interactions among groups, including, for example, exchange of manure from livestock farmers for fodder from arable farmers, and permitted access to crop residues for livestock grazing fodder. Farmers also use personal experience, oral tradition and intergenerational knowledge transfer to inform decisions on crop choice, planting time and cultivation techniques, and grazing regimes (Masters, Baker and Flood 2009). In a changing climate, even under socially and politically stable conditions such decisions become more difficult if not supported by accurate climate predictions and sources of information. Therefore, climate change adaptation involves not just individual farmer responses but also complicated knowledge acquisition and trade-offs among groups built up over many years. These are particularly prone to disruption in post-conflict situations which have broken down social cohesion and perhaps involved displacement.

In other words, the fluidity of population movements is a key issue that needs to be borne in mind in conflict-affected environments, particularly where protracted armed conflicts are bound to mean multiple displacements for populations. Therefore, the concepts of 'community' and 'community structures' often exist only in the imagination of externally funded reconstruction programmes. In reality, many population groups, both urban and rural, try to establish their societal structures in a post-conflict environment and what might have existed before the war in terms of local farming knowledge and techniques, and capacities of dealing with adversaries through family and community structures, may no longer be there.

In post-conflict environments, there is also an assumption that populations would always like to go back to their pre-war lives. However, it is essential to recognize that protracted conflicts do change socio-economic characteristics. A community may have been occupied in agriculture in the past, but having lived in a refugee camp for over two decades, it would be questionable whether they would still have a similar interest to go back to their agricultural communities. This is particularly the case for generations who have been born and bred away from their homes of origin and livelihoods. Similarly, having fought for years in different parts of the country and gained new skills and been exposed to different ways of life, former combatants might not have similar interests to rebuild their lives as farmers. Even if they do, they might no longer have the necessary skills, capacities and knowledge in agriculture. Therefore, in the challenging environments of war-torn societies the climate change would be likely to act as a magnifier of those existing shortcomings. Overall, restoring complex agroecosystems in post-conflict situations, possibly with relocated communities, to a state where they can effectively adapt to climate change presents a significant challenge. This is clearly more difficult than re-establishing conventional agricultural production systems that depend on external inputs typical of more developed countries.

Post-conflict Interventions

Post-conflict recovery and reconstruction in agricultural communities should incorporate measures that are generally considered prerequisite for effective response to climate change in any communities. It is important that both short- and long-term responses to conflict recognize the overarching effects of climate change and build adaptation into the responses so as to maximize the ability of recovering communities to restore adaptability. Conversely, where post-conflict recovery and reconstruction is supported by external aid and development assistance to agriculture, it is important to avoid introducing techniques that reduce the adaptive capacity of communities. However, this could be a difficult challenge, as post-conflict reconstruction is often dictated by a wide range of external and internal agendas, and reconstruction programming according to the specific needs of that particular context may not always be at the top of the agenda. For example, agriculture rehabilitation interventions by different organizations or countries would introduce techniques that might not be suitable for that particular ecosystem. Some agricultural interventions can even damage the possibility of adaptability to climate change, if they create dependency on certain type of seeds, fertilizers and machinery. For example, there are often no controls on the introduction of GM crops in post-conflict environments. Whether or not a particular type of seed or crop should be planted in that particular programme area is often based on the decision of a programme officer. Such decisions are often based on personal experiences, as a particular technique or crop type might have happened to be successful when used by that officer in a different conflict-affected environment.

The adverse effects of climate change can be considered as a 'slow-onset' disaster and adaptation to climate change is usually considered as a medium- to long-term process. In post-conflict situations the priority has to be in dealing with the immediate 'rapid-onset' disaster but steps need to be taken to ensure that issues related to long-term planning for climate change are addressed. Practical Action, a UK-based NGO, has developed an adaptation strategy (Clements, Cossío and Ensor 2010) which would provide a useful framework for assessing the effectiveness of post-conflict interventions in the context of climate change.

The three components of the strategy are vulnerability reduction, resilience and adaptive capacity. Vulnerability reduction encompasses basic development work, often the introduction of specific knowledge or technologies to overcome specific constraints, such as water harvesting or drip irrigation as responses to drought. Resilience is related to the ability to sustain productive capacity in response to climate change by adopting agroecological approaches to the management of soils, crops and livestock. Adaptive capacity refers to the knowledge, ability and self-reliance of people to control and make decisions regarding food production and the use of biodiversity.

In the context of climate change, post-disaster interventions may seek to re-establish the practices in place before conflict that support the resilience and adaptive capacity of agricultural communities. There is also the possibility that post-conflict recovery may provide the opportunity, perhaps by the development of more agroecological approaches, to positively address some aspects of pre-conflict agriculture that increased vulnerability or decreased resilience. The opportunities for addressing climate change in post-conflict agriculture are, to a degree, analogous with the framework to identify the dynamic livelihood strategies and transition processes (Dorward et al. 2005) in which poor households adopt 'hanging in, stepping up or stepping out' livelihood strategies. In the post-conflict situation this equates to re-establishment of the previous agricultural system and protection of best practice for resilience and adaptation to climate change, the establishment of agricultural systems based on the pre-conflict situation but deliberately incorporating new measures to increase self-reliance and reduce vulnerability, and third, to consider more radical, possibly novel, agroecological approaches to address fundamental concerns about long-term response to climate change. However, Smith and Vivekananda (2009) caution that '[t]he top-down imposition of new climate knowledge upon communities is likely to be ignored or received with contempt and reluctance to change'.

In addition to post-conflict intervention affording the opportunity to build in resilience to climatic change, it is also likely that interventions that effectively address climate change issues can reduce future risks of conflicts related to natural resources. Smith and Vivekananda (2009) conclude that this can be achieved by developing integrated natural resource management systems and facilitating communication and cooperation within and between communities, and with government authorities. In a specific example, they consider that reintegrating ex-combatants and resettling displaced persons and refugees in agricultural communities can be more effective and stable if the issues of climate change are addressed so as to ensure economic security of the agricultural system. Such a reform-centred reconstruction approach in order to address existing shortcomings to mitigate future challenges should certainly be an important aspect of peacebuilding operations. However, it is important to find a fine balance between the implementation of reform in the agricultural sector and the existing traditional techniques, methods, structures and beliefs. In most traditional societies, such indigenous structures are imperative for local agricultural production and food security. Taking them away or changing them through reform needs to be gradual and ensure local buy-in, as otherwise this can expose local communities to a new, different set of challenges. A new technique of irrigation or use of a new crop or establishing Western-style cooperatives for marketing to increase the adaptability of communities can also have a set of negative implications. The fragility of environment for agricultural communities in post-conflict environments requires a considered handling of the change process. Western-centric and positivist principles of developmental change may seem to have everything required to prepare communities for the climate change and its impact on the agricultural sector. However, unless they are undertaken in connection with what already exists they will have no ownership of local populations and will not last long.

Conclusion

Climate change as a trigger of future conflicts is likely to be an emerging challenge faced by humanity over the next few decades. As the review of its potential impacts on the agricultural sector shows, future generations will need to deal with the stressors imposed by climate change in terms of a wide range of vulnerabilities. The discussion here focused on climate change as a trigger of new conflicts and a potential factor that could exacerbate post-conflict social vulnerabilities. As far as conflict-affected environments are concerned, the concept of 'starting point' vulnerability should be used as the main approach in the integration of the climate change factor in post-conflict reconstruction of the agricultural sector. As there would be a wide range of processes and stressors in a typical post-conflict environment, social vulnerability would be a particular concern in such a context. Therefore, the reconstruction of the agriculture sector should be in relation to the most effective ways of addressing such sensitivities of the human environment to climate change. Social vulnerabilities that have been exacerbated by the impact of the conflict on societal relationships would be particularly significant in such a post-conflict strategy. To understand the way that war-torn societies can be divided along ethno-religious lines and other socio-cultural and demographic structures and how such divisions could be responded to would be the key starting point in such interventions. In other words, rather than perceiving the post-conflict reconstruction of the agricultural sector solely in terms of natural resources, seeds, fertilizers and farming techniques, the main objective needs to be ways of mainstreaming approaches of social vulnerability reduction into the technical, management and administrative aspects of these interventions.

References

Bruck, T. and Schindler, K. 2009. The impact of violent conflicts on households: what do we know and what should we know about war widows? *Oxford Development Studies*, 37(3), 289–301.

Clements, R., Cossío, M. and Ensor, J. 2010. *Climate Change Adaptation in Peru: The Local Experiences*. Lima: Soluciones Prácticas.

DFID. 2005. *Growth and Poverty Reduction: The Role of Agriculture*. London: DFID.

Dorward, A., Anderson, S., Nava, Y., Pattison, J., Paz, R., Rushto, J. and Sanchez Vera, E. 2005. *A Guide to Indicators and Methods for Assessing the Contribution of Livestock Keeping to the Livelihoods of the Poor*. London: Department of Agricultural Sciences, Imperial College.

Gbetibouo, G.A. and Ringler, C. 2009. *Mapping South African Farming Sector Vulnerability to Climate Change and Variability*. IFPRI Discussion Paper 885. Washington, DC: International Food Policy Research Institute.

Intergovernmental Panel on Climate Change (IPCC). 2007. Summary for policymakers, in *Climate Change 2007: Impacts: Adaptation and Vulnerability. Contribution of Working Group II to the Fourth Assessment Report of the Intergovernmental Panel on Climate Change*, edited by M.L. Parry, O.F. Canziani, J.P. Palutikof, P.J. van der Londen and C.E. Hudson. Cambridge: Cambridge University Press, 7–22.

International Food Policy Research Institute (IFPRI). 2009. *Climate Change Impact on Agriculture and Cost of Adaptation*. Food Policy Report. Washington, DC: IFPRI.

International Food Policy Research Institute (IFPRI). 2010. *Food and Water Security under Global Change: Developing Adaptive Capacity with a Focus on Rural Africa*. CPWF Project Report. Project Number 53. IFPRI and Partners. Available at: http://www.waterandfood.org/uploads/publication_pictures/1288849103_PN53_IFPRI_Project%20Report_Apr10_approved.pdf [accessed: 24 May 2011].

Jones, L., Jaspars, S., Pavanello, S., Ludi, E., Slater, R., Arnall, A., Grist, N. and Mtisi, S. 2010. *Responding to a Changing Climate: Exploring How Disaster Risk Reduction, Social Protection and Livelihoods Approaches Promote Features of Adaptive Capacity*. Working Paper 319. London: Overseas Development Institute.

Lobell, D.B., Burke, M.B., Tebaldi, C., Mastrandrea, M.D., Falcon, W.P. and Naylor, R.L. 2008. Prioritizing climate change adaptation needs for food security in 2030. *Science*, 319, 607–10.

Masters, G., Baker, P. and Flood, J. 2009. *Climate Change and Agricultural Commodities*. CABI Position Paper. Wallingford: CABI.

O'Brien, K., Leichenko, R., Kelkar, U., Venema, H., Aandahl, G., Tompkins, H., Javed, A., Bhadwal, S., Barg, S. and Nygaard, L. 2004. Mapping vulnerability to multiple stressors: climate change and globalization in India. *Global Environmental Change*, 14(4), 303–13.

Ramsbotham, O., Woodhouse, T. and Miall, H. 2011. *Contemporary Conflict Resolution*. Third Edition. London: Pluto Press.

Ringler, C. 2010. Climate change and hunger: Africa's smallholder farmers struggle to adapt. *Eurochoices*, 9(3), 16–21.

Smith, D. and Vivekananda, J. 2009. *Climate Change, Conflict and Fragility: Understanding the Linkages, Shaping Effective Responses*. London: International Alert.

Thomas, D.S.G. and Twyman, C. 2005. Equity and justice in climate change adaptation amongst natural-resource-dependent societies. *Global Environmental Change*, 15, 115–24.

Thornton, P.K., Jones, P.G., Owiyo, T., Kruska, R.L., Herrero, M., Orindi, V., Bhadwal, S., Krisjanson, P., Notenbaert, A., Bekele, N. and Omolo, A. 2008. Climate change and poverty in Africa: mapping hotspots of vulnerability. *AfJARE*, 2(1), 24–44.

PART II
Impact of Conflict on Agriculture

Chapter 4

The Effects of Civil War on Agricultural Development and Rural Livelihood in Sierra Leone

Kwadwo Asenso-Okyere and Sindu Workneh Kebede[1]

Introduction

Sierra Leone, a promising country of 5.8 million people with a land area of 7.2 million hectares (FAO 2009), was hit by a decade-long civil war that started in 1991 and ended in 2001. A mix of political, social, and economic factors within Sierra Leone caused the war. The root causes included widespread poverty, unemployment, low production and productivity, and inequity in access to the nation's resources, which led to pervasive disenchantment of Sierra Leoneans. The war resulted in massive devastation of the physical, social, cultural and economic fabric of the country and caused widespread deprivation and suffering. This chapter is devoted to the effects of the war on the agricultural sector, in which more than 60 per cent of the working people are employed.

Sierra Leone has diverse agroecology, and it is divided into three provinces and 12 districts. About 61 per cent of the population lives in rural areas, although the urban population is increasing at a fast rate of 3.3 per cent compared to 1.9 per cent for the rural population (FAO 2009). The economy of Sierra Leone is predominantly agrarian: agriculture contributed about 51 per cent to gross domestic product (GDP) in 2009 while the industry and service sectors' shares were about 22 per cent and 27 per cent respectively (World Bank 2010).

Poverty and food insecurity form a vicious circle in Sierra Leone. Both afflictions became widespread as agricultural productivity stagnated and the population grew faster than agricultural production. Using the national poverty line of Le 2,111[2] per day, the poverty headcount (P_o) estimates that about 70 per

---

1   We acknowledge the assistance given to us by Professor Edward Rhodes, former director general of the Sierra Leone Agricultural Research Institute, both anecdotal and in terms of data.

2   The national currency of Sierra Leone is the Leone (Le); US$1 = Le 4,300 as of April 2011.

cent of the population of Sierra Leone were poor and 26 per cent were extremely poor in 2005 (PRSP 2005). The poor can meet only about 71 per cent of their basic needs (PRSP 2005). Poverty is heavily concentrated in rural areas and urban areas outside the capital, Freetown. Rural areas contribute almost 73 per cent to the total incidence of poverty. Major causes of poverty are deemed to be multidimensional, including high unemployment, low economic growth and poor social services.

There is widespread food insecurity in Sierra Leone, with 68 per cent of the population not able to afford enough food. The proportion is almost 75 per cent in the provincial states, as compared to 38 per cent in Freetown (PRSP 2005). On the Global Hunger Index (GHI) for 2010,[3] Sierra Leone was among the six worst performers out of 84 countries (von Grebmer et al. 2010).

Agriculture in Sierra Leone

Agriculture is an important sector of the economy of Sierra Leone, contributing 43 per cent in 1995, 59 per cent in 2000, and 51 per cent in 2009 to GDP (World Bank 2010). Out of a total land area of 7.2 million hectares, 5.4 million hectares (75 per cent) is potentially cultivable (IFPRI 2006). Crop production is the dominant activity in agriculture (see Table 4.1), with food crops providing the main source of livelihood for over 75 per cent of the smallholder farmers. This is followed by fisheries, forestry and livestock (MAFFS 2004).

Table 4.1   Percentage contribution to agricultural GDP of subsectors

	1996	2000	2004
Crop	62	66	62
Fisheries	27	21	23
Forestry	6	9	10
Livestock	5	4	5

Source: MAFFS 2004.

---

3   The Global Hunger Index (GHI) is calculated based on three indicators: the proportion of undernourished as a percentage of population, the prevalence of underweight children under the age of five, and the under-five mortality rate (von Grebmer et al. 2010).

Crop Production

Major annual food crops in Sierra Leone are rice, cassava, maize, millet, sorghum, sweet potato and groundnut. Rice is the main staple crop grown by most of the smallholder farmers. In the 1970s, Sierra Leone experienced self-sufficiency in rice with production of over 600,000 tons. However, production dropped in the 1980s, averaging 430,000–524,000 tons per year (FAO 2009). Comparing the rice yield of Sierra Leone with neighbouring countries, we find contrasting results for irrigated and rain-fed rice. The average yield of irrigated rice in Sierra Leone is 2.33 tons per hectare as compared with 3.3 tons per hectare for neighbouring Guinea and 2.91 tons per hectare for the West Africa subregion (IFPRI 2006). However, Sierra Leone does better in yield of rain-fed rice than the average for West Africa and that of neighbouring countries. This implies that Sierra Leone has comparative advantage in producing rain-fed rice in West Africa and this should be exploited as an export crop targeted at the subregion. Continued production of irrigated rice depends upon improvement in yield if Sierra Leone has to compete with other West African countries that produce rice.

The major perennial crops of Sierra Leone are oil palm, citrus, sugar cane, cocoa, coconut and coffee. Oil palm has the highest level of production and is an important part of the upland farming system. Citrus and sugar cane are dominant perennial crops after oil palm. About 87 per cent of the area under cocoa and 73 per cent of the area under coffee could be found in the Eastern Province (FAO 2009).

Fisheries and Livestock

Sierra Leone has coastal and inland waters which provide a wide variety of fish, including high-value species such as shrimps, lobsters, cuttlefish, breams and snappers. The fisheries of Sierra Leone constitute three major sectors, namely artisanal fishery, industrial fishery, and inland fishery and aquaculture.[4] The artisanal fishery operates in estuaries and coastal waters and involves more than 18,000 fishermen who produce mainly for local consumption (MAFFS 2004). Total annual artisanal fish production ranged between 40,000 and 53,000 tons from 1983 to 2002 (MFMR 2002). The industrial fishery, which is export-oriented, operates in open deep waters with vessels and carriers owned by multinational firms in joint venture arrangements with Sierra Leonean nationals. Industrial fish production ranged from 11,000 to 185,000 tons during 1983–2002 (MFMR 2002). The inland fisheries and aquaculture operate in rivers, lakes, floodplains and swamps, with potential yield ranging from 16,000 to 40,000 tons per annum (MFMR 2002).

---

4   Aquaculture refers to the farming of aquatic organisms such as fish under controlled conditions.

The fishery subsector faces constraints in utilizing its full potential. These constraints include a lack of suitable preservation facilities to keep fish fresh, and inadequate road networks for efficient transportation of fish. Inability to meet international standards led to a nine-year ban of fish imports from Sierra Leone into the European Union.

The traditional livestock system in Sierra Leone consists of cattle, small ruminants, pigs and poultry, with cattle dominating. The cattle population, which mostly originate from Guinea-Conakry, is confined to the north of the country and it is largely owned and taken care of by members of the Fula ethnic group. There is no large-scale commercial livestock farming in Sierra Leone and this is largely attributed to lack of financial resources and technical know-how.

Agricultural Marketing

The Sierra Leone Produce Marketing Board (SLPMB) established by the government monopolized the agricultural marketing system (both inputs and outputs) for some time. After liberalization of the marketing system in 1988, small private operators assumed a major role in the marketing of locally produced and imported food commodities.

Perennial crops including cocoa, coffee, oil palm and ginger provide the most important agricultural export commodities of Sierra Leone. Major imports of Sierra Leone include milled paddy rice, maize, onion, vegetable oil and very recently unshelled groundnuts. There is also trade in fish, with both imports and exports. Apart from the formal exports, a lot of informal trade occurs with neighbouring Guinea-Conakry and Liberia, due to bad road networks within Sierra Leone and ethnic and cultural similarities with border towns.

Civil War in Sierra Leone

During the war in the 1990s, the economy of Sierra Leone was volatile and, on average, contracted at a rate of 4.5 per cent per annum (World Bank 2007). GDP growth per capita declined from 2.0 per cent in 1991 to −18.7 per cent in 1992 (see Figure 4.1). Per capita income plummeted by 47 per cent, leading to an exacerbation of poverty, especially in rural areas.

Just before the war, Sierra Leone had a lower poverty rate of 29.8 per cent, as compared with 69.6 per cent for its neighbour Guinea and 54.9 per cent for the average for the West Africa subregion (IFPRI 2006). The poverty situation deteriorated during the war, and by the year 2000 Sierra Leone had among the highest poverty rates in West Africa, 71.8 per cent, compared with 64 per cent for Guinea and 57.8 per cent for the subregion. This is an indication of the devastation and misery the war brought to the country.

From the 84 countries for which the 2010 Global Hunger Index (GHI) was calculated, Sierra Leone was 79th, beating only Ethiopia, Chad, Eritrea, Burundi

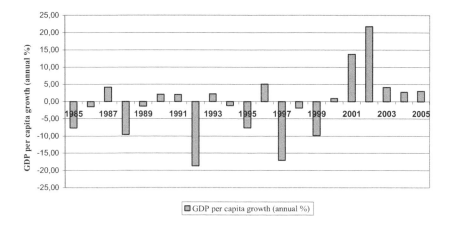

Figure 4.1   Sierra Leone annual GDP per capita growth (%) (1985–2005)

Source: World Bank 2007.

and the Democratic Republic of Congo (IFPRI 2010). Von Grebmer et al. (2008) identified war and violent conflict as the major causes of widespread poverty and food insecurity in most of the countries with high (unfavourable) GHI scores. Even though the GHI score for Sierra Leone in 2010 has improved compared to the 1990 score, the country's hunger status still remains 'alarming' (IFPRI 2010).

The National Nutritional Survey of 1990 indicated a high prevalence of malnutrition in Sierra Leone even before the civil war. About 35 per cent of children had stunted growth, 8 per cent had wasted bodies, and 27 per cent were underweight. Towards the end of the war, in 2000, wasting had increased to 10 per cent, while stunting and underweight remained at the same rate as before the war. The increase in wasting (which points to short-term food insecurity) from 1990 to 2000 can be attributed, among other factors, to the war. Moreover, because the war was fought mostly in the provincial states rather than in the capital, Freetown, about 75 per cent of the population in the provincial states was reported to lack access to adequate amounts of food, as compared to 38 per cent in Freetown.

Effect of the Civil War on Agricultural Production and Marketing

The agricultural sector was heavily hit by the war, which caused a reduction in production. Farm families fled rural areas, abandoning their household belongings and productive assets, including land and livestock. The result was a devastating impact on crop and livestock production.

About 70 per cent of the national livestock herd was destroyed due to the civil war (see Table 4.2). Although the traditional rural smallholder livestock farming system is common in Sierra Leone, peri-urban livestock farming became popular after the war. This is partly because many of the civilians who moved into towns during and after the war settled with their livestock.

Table 4.2   Changes in livestock assets before and after the civil war

	Cattle	Sheep	Goats	Pigs	Poultry (in birds)
1979	333,200	264,000	145,000	17,000	3,000,000
2004	102,000	79,200	43,500	5,100	900,000

Source: MAFFS 2004.

The effect of the civil war on rice production was tremendous. Rice production fell from 503,000 tons at the beginning of the civil war in 1991 to 355,000 tons in 1995 and 360,000 tons in 2001 (see Table 4.3), mainly due to a fall in area planted and harvested. Rice area harvested declined from 372,000 hectares in 1991 to 186,000 hectares in 2001 (FAO 2009). After the civil war, rice production improved due to resettlement of farmers at their farms, which led to a slight increase in area planted and harvested. Consequently, production increased to 542,000 tons in 2004 and to 1,062,000 tons in 2006.

At the beginning of the civil war in 1991, cassava production stood at 123,000 tons, and by the war's end it reached 300,000 tons (Table 4.3). Unlike rice, the civil war did not seem to affect production of cassava, rather cassava production increased due to shifts in food habits and cropping patterns from risky paddy rice to relatively safer crops such as cassava and sweet potatoes. Due to its hardiness and ability to survive under extreme environmental conditions, cassava was regarded as an emergency food. By 2004 annual cassava production was 390,000 tons, and it slightly declined to 350,000 tons in 2006. From 1980 to 2006, yield of cassava ranged between 3 and 6 tons per hectare with an average of 4.9 tons per hectare (FAO 2008). This level of yield is lower than the average for West Africa of 7.7 tons per hectare (IFPRI 2006), implying that there is a need for cassava yields to improve in Sierra Leone to make it competitive in West Africa.

Considering major perennial crops, the war resulted in enormous decline of agricultural production. Oil palm production reduced gradually during the war to 180,000 tons from 250,000 tons in 1991. This was mainly due to a decline in area harvested. Production, however, has not improved after the war; rather it continued to decrease to 174,000 tons in 2004 and to 166,000 tons in 2006. Sugar cane production also declined from 70,000 tons in 1991 to 24,000 tons in 2001. After the war, it picked up to 28,000 tons in 2003 and to 70,000 tons in 2006.

Table 4.3   **Production (×1,000 tons) of major annual and perennial crops (1980–2006)**

	Rice, paddy	Cassava	Maize	Oil palm fruit	Sugar cane	Coffee, green	Cocoa beans
1980	513	95	12	250	NA	10	8
1981	500	97	14	250	50	9	9
1982	523	100	15	250	80	8	15
1983	460	105	15	250	80	16	12
1984	504	100	14	250	70	18	16
1985	430	110	14	250	70	26	18
1986	524	113	8	250	70	23	20
1987	465	114	11	250	70	24	23
1988	493	116	11	250	70	25	23
1989	517	117	11	250	70	25	24
1990	503	123	12	250	70	25	24
1991	503	123	11	250	70	26	24
1992	478	117	10	235	49	26	5
1993	486	105	9	235	24	24	5
1994	405	243	8	231	56	27	11
1995	355	219	8	225	21	25	10
1996	391	281	8	238	21	25	10
1997	411	309	9	245	21	30	13
1998	328	289	8	200	21	26	13
1999	247	239	8	175	21	15	10
2000	199	240	8	175	21	15	10
2001	360	300	10	180	24	17	10
2002	422	340	12	180	24	17	11
2003	445	377	16	195	28	18	12
2004	542	390	32	174	70	16	9
2005	738	390	39	166	70	15	8
2006	1,062	350	48	166	70	15	13

Source: FAO 2008.

Coffee production recorded a decrease during the period of the war. Production declined from 26,000 tons in 1991 to 17,000 tons in 2001. The downward trend in coffee production continued even after the war. Cocoa production reduced sharply during the war from 24,000 tons in 1991 to 10,000 tons in 2001. This was due to huge decline in area harvested from 62,000 hectares before the war to 30,000 hectares after the war. Production gradually picked up during the years after the war to 12,000 tons in 2003 and to 13,000 tons in 2006, but it has not reached its level before the war. Both coffee and cocoa are labour-intensive ventures, and with the desertion of villages by farmers during the war, it was not surprising that production of the two crops went down so much.

A large proportion of the Soviet fleet withdrew in 1991 due to the war, resulting in a drastic decrease in the number of licensed deep fishing vessels. This led to a sharp fall in industrial fish production which remained unchanged until 2011. In contrast, the artisanal fishery did not exhibit much change in production during the war. Before the war started, average fish production from the artisanal fisheries was 45,800 tons (MFMR 2002). During the war period, artisanal fish production ranged between 40,000 and 47,000 tons with an average of 45,900 tons. In the year following the war, production increased to 53,000 tons (MFMR 2002). Even though fishers were reported to have lost fishing boats and fishing equipment during the war, the available catch assessment data for the period did not reflect this.

Agricultural marketing was severely affected by the war. Road infrastructure constraints have limited the area over which private operators carry their business. The road network, which started to deteriorate before the war, was made worse during the war and many food-producing areas became inaccessible. The market structures of many communities were destroyed or left to decay during the war, thereby hampering the sale of foodstuffs. However, food marketing has picked up after the war.

During the years of the war, coffee exports declined sharply from 6,200 tons in 1991 to 1,240 tons in 2001 (FAO 2009). Coffee exports, however, have not picked up after the war. Rather, they declined to 950 tons in 2004. Similarly, cocoa exports registered a decline from 13,000 tons in 1991 to 2,500 tons in 2001 (FAO 2009). The volume of cocoa exports improved after the war to 7,400 tons in 2004. On the other hand, the volume of imported rice, including food aid, increased to make up for the shortfalls in local rice production, putting pressure on the trade balance of the country.

In addition to disruption of international agricultural trade, local markets were massively damaged due to the war. Richards (1996) states that rebels deliberately damaged local markets and social networks through creating fear and suspicion within communities. Nevertheless, despite travel restrictions and trade controls imposed, a marketing opportunity was created by the rebels who had then increased their purchasing power through various nefarious activities. However, only those traders who were willing to take the risks involved in operating in these conditions benefited (Longley, Kamara and Fanthorpe 2003). For a large proportion of the population, there was a decline in livelihood as a result of low production and low income.

Effect of the Civil War on Agricultural Research

Sierra Leone has had a long history of agricultural research, spanning almost 100 years. Centres of activity include: agronomic research at the Njala Experiment Station, Southern Province; Rice Research Station at Rokupr, Northern Province; Veterinary Station at Teko, Kabala, and a Livestock Station at Musaia, both in Northern Province. In 1953, the oil palm research programme at Njala became the West African Institute for Oil Palm Research. From 1953, forestry research was carried out at the Forestry Research Station at Bambawo, Eastern Province. Fisheries research was conducted at the West African Fisheries Research Institute at Kissy near Freetown. The interesting development in the historical account is the number and distribution of research facilities among the various provinces of the country, which attest to the extensive coverage of research conducted in Sierra Leone.

Sierra Leone's promising agricultural research system was destroyed during the civil war. Rebels of the Revolutionary United Front (RUF) and soldiers of the Armed Forces Revolutionary Council (AFRC) entered the premises of the Institute of Agricultural Research (IAR) and the Rice Research Station (RRS) and caused considerable damage to the physical structure, especially that of the RRS. Looting by the Rokupr community continued after the rebels fled the area.

Buildings were destroyed and looting of fixtures, furnishings and equipment was common at the RRS. Some buildings took direct rocket hits; roofing, electrical fittings and lavatories were removed from staff houses and administrative buildings; and furniture was carted away. Research facilities and expensive equipment donated by the International Atomic Energy Agency were destroyed. The seed store/cold room was damaged, resulting in loss of rice and sorghum seed collections and numerous advanced breeding lines. The 150 kilovolt electricity generator, property of the National Power Authority, and two 20 kilovolt standby generators, belonging to the RRS, were damaged, as were transmission lines. Swamps for lowland rice research and the production of breeder and foundation seeds were abandoned as a result of the war; this led to further deterioration of the facilities.

The information available on the effects of the civil war on the human resources of the agricultural research stations is anecdotal, as surveys were not performed. It is, however, sufficient to say that many well-trained Sierra Leonean scientists, including researchers from the university and the research institutes, fled their campuses and took refuge in Freetown or left the country. Those who left Sierra Leone went to Guinea, Gambia, Ghana, Togo, Côte d'Ivoire and other West African countries, and even outside Africa to Europe and North America.

During the war, some shocking events happened to the research system and the educated population of the nation. The first director of the IAR, Professor M.T. Dahniya, and his wife were shot dead by rebels in the outskirts of Freetown. His research coordinator, Professor. J.B. George, had earlier suffered a stroke as a possible consequence of the trauma of the seizure of IAR vehicles by AFRC forces and other stressful events; he subsequently died.

The good news is that most researchers returned after the war. Hence, the negative effect of the war was mainly a loss of research outputs, infrastructure, and the capacity to generate new knowledge and improved agricultural technology rather than a reduction in the number of researchers per se.

Rehabilitation of the Agricultural Sector after the Civil War

The cessation of hostilities in 2001 brought about an improvement in Sierra Leone's security situation and paved the way for economic reconstruction. GDP growth rose from 3.8 per cent in 2000 to 5.4 per cent in 2001, 6.3 per cent in 2002, 7.2 per cent in 2006, and 7.5 per cent in 2007 (MAFFS 2004). This reflected the continuing recovery of agriculture and the expansion of activities in the manufacturing, construction and services sectors. However, according to the 2010 World Development Indicators report, the GDP growth rate has been falling since then, registering 5.5 per cent in 2008 and 4.0 per cent in 2009.

The government of Sierra Leone concentrated its efforts on the resettlement of households displaced by the war in their original communities by providing planting materials, livestock and microcredit schemes (especially for women farmers). In collaboration with donors, an interim poverty reduction strategy paper (PRSP) and an action plan were developed in 2001 that focused on improving the livelihood of the people of Sierra Leone. In addition, an Agricultural Sector Review and Strategy for Development was developed in 2004 to supplement the PRSP.

Among the rehabilitation efforts in the agricultural sector, the fisheries subsector has gone a long way to resettle fishermen whose livelihoods were disrupted due to the war. Apart from the emergency programmes, the sector benefited from funds generated from the Highly Indebted Poor Country (HIPC) initiative in the form of provision of fishing inputs and the construction of efficient ovens in many coastal fishing communities. In addition, the Artisanal Fisheries Development Project (AFDP) provided complete fishing tools to cooperative members in most coastal districts and trained fish processors in fish handling and processing techniques. According to the report of Fisheries of Sierra Leone (2002), fish exports after the war period increased slowly from 1,025 tons in 2002 to 1,800 tons in 2007 (MFMR 2002).

There has been a revival of Sierra Leone's ginger industry, which was seriously hit by the war. An initiative by the International Trade Centre (ITC) provided work and income for more than 9,000 farmers, especially women. The peeling, drying and processing of ginger increased its value by about 90 per cent and the product was exported to Europe and India. According to the ITC (_International Trade Forum_ 2007), Sierra Leone exported a total of 4 tons of ginger to 11 countries in 2006, and planned to export 80 tons of dried ginger in 2007. This shows how special initiatives can provide employment and income for disadvantaged persons in unusual circumstances.

In terms of rebuilding the agricultural research system after the war, the Sierra Leone Agricultural Research Institute (SLARI) was established in 2007 by an Act of Parliament (Government of Sierra Leone 2007). SLARI was made the main agricultural research and agricultural technology-generating body for the benefit of the farming, fishing and forestry sectors in Sierra Leone and to provide for other related matters. When fully operational, SLARI would have eight research centres in different parts of the country to generate new knowledge to improve agricultural production and productivity in Sierra Leone. To provide a clear sense of direction and make an impact, SLARI developed a strategic plan and an operational plan in 2008. However, for effective dissemination of knowledge among farmers and other users along the value chain there is the need to overhaul the agricultural extension and advisory services alongside the revitalization of agricultural research.

Agricultural Growth Prospects for Sierra Leone

The agricultural development strategy of Sierra Leone focuses on promoting food security and poverty alleviation and making agriculture the 'engine' of growth and development through commercialization of agriculture (Sesay 2007). It is estimated that the share of households with adequate food consumption increased from 56 per cent in 2005 to 71 per cent in 2007, largely as a result of steady increase in domestic production of the major staple crops (Sesay 2008). Specifically, rice self-sufficiency rose from 57 per cent in 2002, to 59 per cent in 2003, to 69 per cent in 2005, and to 71 per cent in 2007. Although Sierra Leone has a comparative advantage in the production of rain-fed rice it is dangerous to rely on rainfall at a time when there is so much uncertainty in rainfall patterns due to climate change. Attention should be paid to raising the yields of irrigated rice to make it profitable so that it can compete in West Africa.

Tree crops have a huge potential for export and foreign exchange earnings in the country. The main crops are cocoa, coffee, oil palm and kola nut, followed by rubber and cashew. The port of Freetown, being an important centre of trade for many countries, provides a huge potential for Sierra Leone's international trade, especially regional trade in West Africa and the rest of Africa. Hence, potential needs to be exploited.

In relation to the fishery subsector, there is an untapped potential that the country can make use of. The total potential sustainable yield from marine fisheries capture is estimated at 112,000–180,000 tons while the current total production stands at an average of 67,000 tons per annum (MFMR 2002). This shows a gap of 45,000 tons of untapped fish resources that can be harvested per annum. Greater attention and effort are needed for the efficient management and sustainable development of fisheries. Moreover, it is important to find ways of improving fisheries marketing for both local and international markets. In this respect, attention should be paid to sanitary conditions so that Sierra Leonean fishery products can qualify for all markets in the world.

There is also an unmet demand for livestock products, as indicated by huge imports of animal products. It is, therefore, important to encourage and promote local production of livestock products to satisfy the domestic market and possibly for export to the regional market. Poultry and pigs, for which the country has a comparative advantage, could be quickly used as an entry point.

Conclusion

Agriculture is the dominant sector in Sierra Leone. Food crop farming practised by smallholders who cultivate between 0.5 and 2 hectares is largely rain-fed. Rice is the major staple crop cultivated by almost every farm household. Other important food crops are cassava, sorghum, maize, millet, groundnuts and sweet potato. Production of rice declined during the war, mainly due to a fall in area planted and harvested. However, rice production increased after the war and this has to be sustained. If this is not done, food security will be greatly compromised in the medium to long term.

The civil war in Sierra Leone has devastated the country's infrastructure (especially road networks), institutions (both formal and traditional), and social networks and structures, among other resources. This had an adverse effect on both domestic and international agricultural markets. For instance, due to bad road networks, farmers close to the national border find it easier to transport and sell products in markets of neighbouring countries instead of distant market centres in Sierra Leone. This creates a demand gap and leaves the country to resort to food imports with their adverse impact on balance of payments and the foreign exchange situation. Sierra Leone needs to improve upon domestic marketing of food, with massive rehabilitation and construction of road infrastructure.

Exports from Sierra Leone have a narrow base of agricultural products and minerals. Most of these commodities are exported as raw materials. It is important to note here that processing the raw materials would create value addition, employment, and increase the country's foreign exchange earnings. As it stands, volumes of production of most of the export items declined sharply during the war and they are still recovering. The government should accelerate the process by undertaking export recovery programmes for agricultural commodities in which the country has comparative advantage such as cocoa and coffee.

In addition, Sierra Leone has considerable potential in the fishery and livestock subsectors. For these, the government needs to provide targeted assistance. In the livestock subsector, government can promote poultry and pig farming to empower small-scale farmers in areas with high population densities. Under the prevailing conditions, it is recommended that farmers should be encouraged to engage in this activity since it does not require large tracts of land and the output per land unit is high. In addition, it is possible to mobilize resources quickly for poultry and pig projects as they have short production cycles and provide the much required employment and meat products, which are in high demand in Sierra Leone.

It must be noted, however, that with 65 per cent of Sierra Leoneans being Muslims, pork may not be part of their diet. Hence, care must be taken about the extent of developing pig production for domestic consumption.

Sierra Leone has potential for growth given its vast resources. The will to succeed has been demonstrated by various post-war recovery efforts including the building of a stable but still fragile economy. Current opportunities and challenges arising from the worldwide high food prices point to increased attention to agriculture as the main source of growth and poverty reduction. Finally, based on the experience of post-conflict recovery of the agricultural sector in Sierra Leone, the following points are crucial as lessons learned.

Lessons Learned for Post-conflict Recovery of the Agricultural Sector:

1. Rebuild institutions (public, civil society including farmer organizations, and private) to spearhead the post-conflict recovery. A plan should be developed for the recovery of the economy in general, and agriculture in particular.
2. People who left their lands to escape the war have to be brought back to restart their farms. This requires psychological preparation as some of them may have witnessed some of the atrocities of the war. Start-up capital must be granted for acquisition of tools of trade and to re-establish livelihood.
3. Immediately after resettlement, attention should be paid to the production of staple crops so that food security is not further compromised. High-yielding varieties of planting materials, livestock or fingerlings should be supplied to farmers and fisherfolk, and rural services for inputs and credit should be established to cater for the needs of producers. As production for household consumption stabilizes, attention should be paid to market-oriented production for both local and export markets to earn income to cater for other household purchases.
4. Investments should be encouraged in areas that provide quick wins, for example short-cycle livestock like poultry and pig production and short-duration crops and foodstuffs for domestic consumption.
5. Markets should be developed to encourage increased production.
6. Extension and advisory services have to be scaled up to speed up the process of adjustment in terms of knowledge acquisition and innovation. There may be the need to employ more extension staff and develop their capacity, and also use more effective extension methods, such as mobile phones and radio to relay information to farmers.
7. The national agricultural research system should be revitalized to generate new knowledge and address farmers' problems.
8. The capacity of the farmers to absorb new knowledge and to demand the knowledge and information they require should be developed.

9. Rebuild infrastructure (especially roads, water, power, schools and clinics) that was destroyed during the war and build new infrastructure to support the recovery.
10. Develop confidence both internally and externally for private investments in the general economy and the agricultural sector.
11. The root causes of the conflict – which included low agricultural productivity and production which contributed to pervasive poverty in the rural areas, and lack of rural amenities which encouraged migration of young people to the urban areas and contributed to shortage of agricultural labour – have to be tackled so that the conflict does not resurface.

References

Food and Agricultural Organization of the United Nations (FAO). 2008. FAOSTAT. Rome: Food and Agricultural Organization of the United Nations.

Food and Agricultural Organization of the United Nations (FAO). 2009. FAOSTAT. Rome: Food and Agricultural Organization of the United Nations.

Government of Sierra Leone. 2007. SLARI Act. Freetown: Government of Sierra Leone.

International Food Policy Research Institute (IFPRI). 2006. *Regional Strategic Alternatives for Agriculture-led Growth and Poverty Reduction in West Africa.* Washington, DC: International Food Policy Research Institute.

International Food Policy Research Institute (IFPRI). 2010. Global Hunger Index. Washington, DC: International Food Policy Research Institute.

International Trade Forum. 2007. In Sierra Leone, ginger trade helps recovery. *International Trade Forum* [Online], January 2007. Available at: http://www.tradeforum.org/news/fullstory.php/aid/1131/In_Sierra_Leone,_Ginger_Trade_Helps_Recovery.html [accessed: 13 June 2011].

Longley, C., Kamara, V. and Fanthorpe, R. 2003. *Rural Livelihoods in Kambia District, Sierra Leone: The Impacts of Conflict, ODI Working Paper* 186. London: Overseas Development Institute.

Ministry of Agriculture, Forestry and Food Security (MAFFS). 2004. *Agricultural Sector Review and Agricultural Development Strategy*, vol. 2. Freetown: Ministry of Agriculture, Forestry and Food Security.

Ministry of Fisheries and Marine Resources (MFMR). 2003. *Fisheries Policy Paper*. Freetown: Ministry of Fisheries and Marine Resources.

Poverty Reduction Strategy Paper Status Report on Sierra Leone (PRSP). 2005. Paper presented at a conference on the Millennium Development Goals (MDGs) in West Africa, Dakar, Senegal, 26–28 February 2003.

Richards, P. 1996. *Fighting for the Rain Forest: War, Youth and Resources in Sierra Leone.* Oxford: James Currey.

Sesay, S. 2007. Agriculture sector coordination mechanisms. Freetown: Ministry of Agriculture, Forestry and Food Security.

Sesay, S. 2008. Synopsis, vision and strategy for agriculture sector development in Sierra Leone. Freetown: Ministry of Agriculture, Forestry and Food Security.

von Grebmer, K., Fritschel, H., Nestorova, B., Olofinbiyi, T., Pandya-Lorch, R. and Yohannes, Y. 2008. *Global Hunger Index: The Challenge of Hunger 2008*. Washington, DC: International Food Policy Research Institute.

von Grebmer, K., Ruel, M., Menon, P., Nestorava, B., Olofinbiyi, T., Fritschel, H., Yohannes, Y., von Oppeln, C., Towey, O., Golden, K. and Thompson, J. 2010. *2010 Global Hunger Index: The Challenge of Hunger: Focus on the Crisis of Child Undernutrition*. Washington, DC: International Food Policy Research Institute.

World Bank. 2007. World Development Indicators 2007. Washington, DC: World Bank.

World Bank. 2010. World Development Indicators 2010. Washington, DC: World Bank.

Chapter 5

How Conflict Affected Agriculture in Nepal

Gobinda Neupane and Richard Slade[1]

Introduction

This chapter provides a case study of the impact of conflict between government and Maoist forces on Nepalese agriculture between 1996 and 2006. Against the key themes of the environment, society, the economy and governance, the chapter aims to explore the impact of hostility on long-established traditional farming practices, developed to deal with a varied and challenging geographical environment. A feature of the study is that the conflict saw a significant deterioration in agricultural production, its associated infrastructure and the socio-economic profile of farming communities, culminating in food shortages in some areas. However, these changes drew on longer-term problems associated with Nepalese agricultural practice. These problems included the environmental challenges of farming in a varied geographical landscape, the societal and economic implications of failing to match population growth with agricultural production, and the lack of capacity in governance of resourcing and implementing agricultural research and technology strategies. Underpinning these difficulties were cross-cutting long-term issues arising from landownership and a steady migration, especially of young people, away from rural communities. The interaction of these background pressures with a decade of hostility was not marked by any specific conflict event. Rather, the effects of the one flowed into the other in a way that has compounded the challenges that continue to be faced by post-conflict Nepalese farming.

Agriculture in Nepal

Nepal's economy has historically been dominated by its agricultural sector, contributing over 60 per cent of GDP until the mid-1980s. Despite the disabling effect of continuing decline and a decade of conflict, in 2007 agriculture still accounted for 33 per cent of GDP (Jones 2010). The continuing importance of

1   The authors would like to acknowledge Dr Rebecca Roberts for her encouragement and support. They would also like to express their thanks to Mr Durga Pd Adhikari, Coordinator SEAN for his support, Mr Bimal Thapa, Department of Agriculture, Mr Padam Bhandari, SNV and Mr Ananta Ghimire for their contribution and advice. Last but not least thanks goes to Mr Pushpa Raj Shrestha, WFP Nepal for providing references.

farming is explained, in part, by the range of altitudes and climatic regions from which Nepal benefits. The country enjoys temperate, sub-tropical and tropical climates, all occurring within a 200 Kilometre north-south distance. Table 5.1 illustrates the relationship of the countries five geographical regions to altitude, climate and type of agricultural produce.

Table 5.1   Geography and agriculture in Nepal

Geographical region	Altitude	Climate	Agricultural produce
Terai	78m–500m	Tropical climate	Cereal crops are paddy, wheat, maize and lentil. Fruit crops-mango, litchi, banana, guava and pineapple. Livestock-buffalo, cow, goat, chicken, ducks
Churia or Siwalik	500m–1000m	Sub-tropical climate	Cereal crops are paddy, maize, millet, wheat, black gram, zinger, turmeric. Fruits are mango, litchi, jackfruit, guava, lime, avocado, sweet orange, mandarin etc. Livestock- cow, buffalo, goat, chicken, pigs
Mid-hills	1000m–2000m	Mild temperate zone	Cereal crops are maize, millet, buck wheat, wheat, paddy, potato, ginger, garlic, cabbage, cauliflower, radish etc. Fruits are citrus, pear, plum, walnut and low chilling varieties of apple. Livestock- cow, goat, pigs, buffalo, sheep
Middle Mountains	2000m–and above	Temperate climate	Cereals crops are wheat, buck wheat, oat, maize, potato, barley. Fruits are apple, walnut, pear only temperate fruits. Livestock- sheep, Chauri (local breed of cow), mountain goat
High mountain	5000m		Not feasible to cultivate- snow land

Table 5.1 demonstrates the crucial diversity of Nepalese farming. This diversity allows for flexibility in responding to climatic or environmental pressures affecting produce in one part of the country, with foodstuffs from another region. Against this geographical background, Nepal's had previously maintained agricultural self-sufficiency achieved through long established farming techniques interacting sympathetically and productively with Nepal's varied landscape and environment. A unique aspect of Nepalese farming agriculture practice is the hilly and mountainous region that has traditionally seen the cultivation of crops on terraces; a technique rooted in the country's ancestral heritage. Indeed the region illustrates the chapter's focus on the relationship between food shortages, failure to modernize agricultural practice, a consequent decline in rural prosperity and inevitable outward migration: and all this before internal hostility broke out.

Nepal has traditionally practised a system of sustainable farming. Measures including mixed cropping – growing different crops such as wheat and mustard at the same time in the same plot, so one can be harvested even in adverse weather conditions – have acted to increase the variety and viability of production. Furthermore, if one crop failed as a consequence of natural disasters, for example drought, hailstones or excessive rain, then practices such as relay or sequential crop growing (or intercropping), involving a second crop being grown without harvesting the first, have been able to ensure a substitute product to sustain communities. This approach to sustainability in Nepalese farming has also seen the close integration of livestock, forest, trees and crop systems. Livestock provides manure for field crops, leading to both produce gain and by-products such as straw used to feed animals. Likewise, forests provide fodder for animals resulting in manure that is returned to the fields, thereby contributing to improved crop harvests.

Within the mid-hill and mountainous regions, farmers have traditionally adopted practices that paid great heed to the care and management of soil and water resources. These included building and maintaining protective stone walls, managing outlet and drainage systems, planting fodder trees and using forest leaves as mulching materials.[2] Manure prepared from cow dung and animal bedding materials has traditionally been used to nourish the land before crop seeds are planted. Arguably, these traditional approaches represented organic natural farming in its truest sense. Although such practices have since been replaced by inorganic fertilizer and pesticide, other long-established farming systems hold fast. For example, there is no scope for mechanization in the hilly areas and as a consequence animal and human power has been the mainstay of agricultural work. Traditionally there have always been sufficient population levels available to provide labour resources in these regions.

In the round, this sustainability was maintained by managing a careful balance of human demands, environmental factors and the integration of key agricultural

---

2   Forest leaves and twigs are used to protect soil moisture covering the field surface during the dry season. They also protect against 'soil splash' from raindrops during the rainy season.

systems. If one component weakened or became disturbed then it would be likely to have a significant impact on the total farming system, overall production and the ability of rural communities to maintain economic viability. Conflict between government and Maoist forces had just such an impact: not on one component but rather across the whole agricultural framework. Yet the systemic failure which accompanied the hostility was itself rooted in long-standing difficulties plaguing Nepalese agriculture and society more broadly and which formed a significant cause of the conflict.

The Rural Background to the Conflict

Given the proportion of the population engaged in agricultural production, the economic and societal conditions that formed a backcloth to the hostility were clearly significant for rural areas. Some of the root causes of the clash can be seen in high levels of poverty and societal exclusion. According to the United Nations Human Development Index (HDI 2010) Nepal ranks 138th out of 162 countries, suggesting high levels of inequality. Whilst poverty levels fell from 42 per cent to 31 per cent between 1995 and 2004 (United Nations Human Development Index 2010), this improvement arose mainly through young people working abroad and sending their income home as remittances. Despite these gains, Nepal lags behind other countries in the region as a result of both political instability and a failure to adopt more liberal economic policies. Together, these factors have restricted inclusive growth.

The broad-based impact of these circumstances across the majority of the population meant that conflict spread throughout the country and affected most people and most communities. Indeed, during the conflict Maoists claimed that the majority of rural Nepal, approximately 80 per cent, was under their control. Although this claim was not independently verified, it was the case that any government presence was limited to their headquarters in districts and cities and that during the decade of conflict most local government offices were destroyed, almost paralysing service delivery. This had a significant impact. It saw the accelerating deterioration of the agricultural system, the breakdown of government initiatives already struggling to realize the benefits of research and new technology, and a growth in established patterns of migration away from rural areas.

Agriculture before and during the Conflict

Agriculture is a major sector of the Nepalese economy, providing 66 per cent of employment opportunities and contributing 36 per cent to GDP in 2010–11 (MOAC undated). As recently as 1990, more than 90 per cent of the population were engaged in the agricultural sector, and in the same year Nepal achieved self-sufficiency in food production. However, even before the conflict broke out there were signs that not all was well. Population increases had led to the need for food being imported, primarily because farming practices were failing to match the demands of demographic growth, a trend that has continued since hostilities ceased and which remains a pressing concern. In 2008 agriculture expansion was recorded at only 0.7 per cent whilst the same year saw the population increase by 2.25 per cent (Ministry of Agriculture 2009). What factors lay behind this long-term failure to match demand and what impact did the conflict have on this situation?

Concerns regarding landownership, especially absentee landlords and migration away from rural areas (the significance of migration is explored later in the chapter), are closely connected and enduring issues in Nepal. Farmers, whose income relies on the land, especially for crop cultivation, are often not the landowners. Of those who do own their land, many hold less than a hectare, giving limited scope for a viable return on their investment of time and effort. Typically, such farmers have no other occupation, and see opportunities to move to city areas as a way of advancing their prosperity. In doing so, they often prefer to retain landownership, both as a fixed asset and to enhance feelings of economic security. This trend increases the number of absentee landlords and affects the capacity of those rural families who choose not to migrate, acquiring sufficient land to generate an adequate income.

This situation is made worse by the extent to which large areas of land were already owned or have been acquired by elites. As a consequence, landless families can be seen to have taken possession of forest areas or government land in order to generate a livelihood. Known as the sukumbasi (landless people), the plight of this group assumed great significance in the conflict. Maoist Party cadres seized land owned by elites who were considered both corrupt and opposed to what Maoist forces saw as a 'people's' war. The Maoists either began farming these captured assets communally or made them available to landless families. However, both scenarios saw productivity reduced, primarily because neither Maoists nor the sukumbasi had expertise in improved farming practices and crucially they were unable to access essential support including technology and infrastructure systems such as irrigation facilities.

The impact of the consequent reduction in agricultural productivity was exacerbated by two other factors restricting the amount of land available for viable agriculture. First, despite a long-established dependency on farming, agricultural land has increasingly been turned over to residential use. This is especially so in Kathmandu, an area known for its fertility, but also applies to agricultural land around most cities. Second, the amount of land available for viable farming is limited by land fragmentation. Traditionally, when young people became independent from their parents they received land on which they could pursue an agricultural livelihood. However, as farming assets were redistributed among family members, the land available to a family became so small – an average family having less than 0.2 hectares to farm – that it proved insufficient for commercial farming.

Two further factors came to bear on this situation. First, before the conflict agro-base industries[3] had proved to be productive. However, this was only because it became more cost-effective to import raw materials rather than rely on indigenous products, pointing to long-standing problems with inadequate infrastructure arrangements for communication, irrigation, marketing and storage facilities and agricultural 'inputs', for example fertilizers. As recently as 2008 the Ministry of Agriculture had identified the poor supply of these 'inputs' as a significant and continuing problem for farmers. However, during the conflict the impact of these difficulties became much more serious. Weak law and order, market disturbance and the blockage of essential road networks disabled the capacity of key government bodies such as the Agriculture Input Corporation to provide essential support, especially to remote areas. The result was that poor-quality fertilizers and pesticides were dumped in the countryside, leaving farmers less productive and as a consequence their communities more deprived. Second, access to essential loan finance to support farming was restricted. Before the conflict, Nepalese farmers had been able to receive loans through the Agriculture Development Bank, although an interest rate of 14 per cent was hardly favourable when compared with other business sectors. However, during the hostility banks in rural areas were unable to operate for security reason, leaving farmers with greatly reduced access to essential finance. Unfortunately the situation is still not much improved as banks have yet to resume a presence in rural areas.

Whilst the supply and ownership of land and inadequate infrastructure systems represented significant problems for rural communities, two fundamental issues underpinned a deep malaise in Nepalese agriculture which the conflict significantly aggravated. First, the failure of agricultural systems to modernize farming practices, and second the movement of agricultural workers, in a labour-intensive industry, away from the countryside.

Modernizing Farming Practices

The Terai region is particularly important for Nepalese agriculture, spanning mountain, mid-hill and plain areas, the latter providing a tropical climate capable of harvesting three annual crops. This contrasts with the mid-hills area where two crops, maize and millet in the uplands and paddy and wheat in lower-lying irrigated land, are produced annually. In the mountains the colder climate sees the principal crops of wheat, barley, potato and buckwheat produced only once each year, giving rise to the danger of local food shortages and reduced prosperity. However, in general Terai is a fertile area producing crops to feed people living in the plains and address food deficits in hill and mountain populations.

---

3   Agro-base industries are those that need raw materials to manufacture their products, for example apples, pears and grapes required to produce jam, or sugar cane required in producing sugar.

The deficits in hill and mountain communities have complex environmental and associated societal causes. Food deficit areas are more prone to natural calamities like drought, hailstones and windstorms, damaging crops and leading to food shortages. As a consequence food-for-work programmes have been implemented in an effort to increase access to food. However, an unintended consequence of such initiatives has been a growing trend in rural populations to prefer to work on these programmes, which provide immediate food to feed families, rather than cultivating more risky areas, where natural calamities see farmers being unable to rely on a product until a harvest is gathered and stored. In an effort to remedy the situation the pre-conflict Nepalese government provided a food transportation subsidy to tackle food shortages in remote hill and mountain districts, supported by a system of government-established food depots. As a consequence of these arrangements people changed their food consumption patterns, with rice becoming preferred to indigenous produce. Arguably both programmes undermined a much needed focus on improving local agricultural practice. More significantly, during the conflict insecurity and associated transportation difficulties saw both forms of support, on which communities had come to depend, function increasingly ineffectively.

Before the conflict, and in a further attempt to remedy growing concerns surrounding food deficits, the Ministry of Agriculture and Cooperatives put in place a programme aiming to improve agricultural technology through developing the skills and knowledge of farmers and growers. Known as the Extension System, this strategy formed a key part of the Ministry of Agriculture's Prospective Plan drawn up in 1995 designed to improve farming practices. However, there were problems with the approach from the beginning. The government decided that services set out in the plan would only be available on demand. Since the majority of the farmers did not know about either the plan or the extension policy objectives, dissemination of knowledge designed to improve agricultural technology was limited. Furthermore, implementation was hindered by the failure of resources to be tailored to the life of the plan. For example, donors such as the UK's Department for International Development (DFID) would only commit resources for a period of three years (MOAC undated), although the plan covered a twenty-year implementation cycle.

From this less than promising beginning, implementation difficulties were worsened further by the internal conflict. Hostilities saw the government realize very limited access to agricultural regions to support farmers through implementing the extension system. Warring Maoist parties suspected any government staff present in rural localities as being either a spy or some other kind of belligerent. Indeed, government staff were often threatened and required to make a 'donation' to the Maoist combatants. Furthermore, agriculture service centres and offices were destroyed with the result that staff increasingly remained in their district headquarters, limiting their service to farmers. This was particularly apparent in government initiatives to provide agricultural research support. Against a background of low-level government resource allocation, most government research farms were affected during the conflict, some being set on fire or otherwise destroyed, preventing these key establishments from operating. This was significant for implementing newly-developed technology, for example to verify the condition of fields. In general, researchers could not develop technological solutions for agriculture problems, resulting in even lower growth for the farming sector as a whole. This deterioration was marked by the inability of the government to maintain and provide growers with foundation seeds.[4]

Nepal has seen a number of efforts made to promote seed production programmes based on cereal and vegetable crops. Hybrid seed production has not been a priority of either government, farmers or the private sector. Instead initiatives have concentrated primarily on open-pollinated crop varieties,[5] with climatic and geographical variation seeing greater scope for the production of vegetable crop seeds. The government's Food and Agriculture Organization (FAO) funded the Fresh Vegetable and Seed Production Programme in 1980, followed by a number of other programmes supported by donors and prioritized by the Department of Agriculture. These initiatives saw seed production programmes operating successfully in some pocket areas, leading to an expansion of such programmes, funded through different donor projects such as the Seed Sector Support Programme. This in turn led to success in creating a network of seed vendors, helping to establish a market for crop seeds. Building on this, farmers from a number of the initial seed production pocket areas saw lucrative business opportunities and started producing crop seeds on a larger scale. Farming began to realize increased profit levels for rural communities. This increased profitability drew on the more advanced knowledge of farmers in seed production technology and the assurance of a market to buy the seeds they produced. Both factors led farmers to start producing seeds primarily for vegetable crops such as radish, carrot, onion, cauliflower, cabbage, beans, tomato and aubergine. In turn, farmers using the seeds found their resulting income was almost double that from cereal crop production.

However, the internal hostility led to local crop varieties being dramatically affected. The lack of any programme to protect local crop varieties saw farmers lose hope in the value of cultivating this kind of produce. More significantly for the future of Nepalese agriculture, the whole seed production system, which had been making such progress, was seriously affected by the conflict. Lack of support from technicians, and the incapacitation of government agriculture farms resulted in a shortage of foundation seeds. This, combined with a failure to achieve a timely supply of agricultural inputs due to a high level of strikes and road closures and the inability of project staff to support existing projects and transfer better technology to new areas, saw seed production, the supply system, and farmers' income collapse. During the period between 1990 and 2000, Nepal had been an exporter of vegetable seeds to Bangladesh, India and other countries. However, the

---

4 Foundation seeds are those certified as complying with high standards of genetic purity and production.

5 Open-pollinated crop seeds are those that are naturally produced without any human interference or control.

downturn in seed farming since 2000 has seen this scenario reversed. Nepal has become a vegetable seed importer, including of hybrid seeds. This approach runs contrary to pre-conflict strategy, has required a high amount of foreign currency and has seen commercial vegetable production increasingly favour hybrid seeds, with expansion only limited by access to roads and local market areas.

According to the Coordinator of the Seed Entrepreneur Association of Nepal (SEAN), some four years after the peace accord, indigenous seed production is only now starting to see progress and has yet to reach the level of earlier stages of production. Progress is still hampered by a lack of foundation seeds from government agriculture farms. Although SEAN has started its own foundation seed production, and put in place supply systems to support seed growers, in general farmers are unable to obtain the volume and quality of crop seeds necessary for good crop production. In order to address this situation, and as evidence of the importance of the programme, the government of Nepal has enacted legislation authorizing the National Seed Board to ensure the private sector develops and maintains an adequate supply of foundation seeds.[6]

Agricultural Labour

Alongside the problematic circumstances surrounding modernizing initiatives, persistent shortages of essential agricultural labour became much more serious during the hostility. We have already explained that Nepal's economy has historically been dominated by its agriculture sector. Although farming had slumped from 60 per cent of GDP in the mid-1980s to only 33 per cent by 2007, agriculture still accounted for 85 per cent of employment in rural areas, contributing around 55 per cent to rural household income (Jones 2010: 6). However, viable employment opportunities in rural areas are particularly lacking due to the sluggish growth of agriculture. Many people attempt both self-employed farming and paid agricultural work but are unable to achieve a guaranteed livelihood, largely because of the deterioration in agriculture generally and the inconsistency and uncertainties inherent in seasonal work.

Understandably, people have sought better options. In the poorest mid- and far west development regions inhabitants have traditionally travelled to India for seasonal labour work. According to the Asian Development Bank, by 2010 the exodus from Nepal involving younger people stood at 45 per cent (Jones et al. 2010: 4–5). For some young people migration has become an important source of income. During the conflict period the mid- and far west regions, identified as localities with high levels of overall need and growing patterns of migration, were significantly affected. People were often obliged to join the Maoist or government warring parties, or support either force in some other way. The only alternative was for people to leave their communities. Given the historical trend towards

---

6   Nepal Seed Act 2045 (B.S.) 1989 as amended in 2008.

employment outside Nepal, migration to India, or city areas within Nepal, became a crucial option for escaping the internal armed conflict.

As a result agricultural labour resources have become seriously depleted. During the decade of hostilities and in the post-conflict period younger people working away from their communities in both skilled and unskilled work saw their minimum earnings stand at about US$200 per month. Even given the essential migration costs of tickets and other expenses, a farming business in their home community could not hope to compete with this level of income. Recent statistics suggest that for 2.1 million younger people the investment in migration costs has been an acceptable expenditure in their search for work away from rural areas (World Bank 2010: i). Furthermore, remittances from those members of the labour force outside the country have become important in sustaining Nepal's economy, underlining the broader acceptance in society of the need to travel in order to secure employment and income.

The new life that migrants have opted for has not been easy. It sees them engaged in a struggle to find employment or establish small businesses in their new domiciles. However, when compared to the harder life of farming in remote districts they have preferred to stay in city areas and continue struggling for a better future. Only a limited proportion of the workforce, unable to realize the opportunities that migration might have brought, have returned home and become involved in farming.

Overall, the critical labour shortage in farming is becoming evident in a number of ways. First, farmers have begun to focus on growing major crops such as paddy, wheat, maize and millet, reducing the intensity and overall levels of agriculture production. Second, some crop varieties require specialist techniques, especially when they involve growing and maintaining seeds. Traditionally, male members of farming families have been the principal decision-makers regarding such matters. Their absence through migration has led to a reduction in available skills and knowledge regarding crop varieties and cultivation. Third, in the post-conflict environment there has been a significant reduction in local crop varieties. This may be hampering biodiversity, leading to increased problems arising from disease and pests. Above all, and despite the peace accord, some rural communities face a vicious circle revolving around migration and a continuing and growing labour shortage. The movement of young people reduces labour resources and thereby the extent to which agricultural decline can be reversed. This directly affects the likelihood of improved farming production and rural prosperity. The outcome is even greater migration.

Conclusion

Clearly much remains to be done to regenerate Nepalese agriculture. In little more than three decades Nepal has moved from a position of food self-sufficiency to one of dependency on food imports and donors in order to meet the needs of its growing population. A failure to improve the efficiency of agricultural production has seen a decline in rural prosperity which, together with inequality and exclusion, contributed to the causes of a decade of internal conflict that devastated infrastructure systems and halted some signs of improvement. However, the peace accord provides the opportunity to reverse the decline. Whilst land reform may provide for farms of more viable size, and improved infrastructure can deliver crucial access to markets and the supply of agricultural inputs, the key issue is to address declining agricultural productivity that has resulted in low levels of income and high levels of migration. In the past attempts have been made to bring farming practices up to date, with some signs of success, before progress was halted by hostility. This argues that a properly resourced and thoroughly implemented modernization strategy may be capable of addressing the low levels of rural prosperity that contributed to the conflict. In doing so, trends of rising migration can be reduced, thereby helping communities regain viability, enabling Nepal ultimately to approach food self-sufficiency once again.

References

Asian Development Bank. 2009. *Nepal: Critical Development Constraints*. Available at: http://www.adb.org/Documents/Studies/Nepal-Critical-Develop ment-Constraints/Main-Report-Nepal-Critical-Development-Constraints/ default.asp.

Jones, S. 2010. *Policymaking During Political Transition in Nepal*, Oxford Policy Management Working Paper 2010-03. DFID.

Jones, S. et al. 2010. *Defining the Political Opportunities and Constraints in Key Economy Sectors for Promoting Inclusive Growth*. World Bank and DFID.

Ministry of Agriculture, Nepal. 1991. *Agriculture Prospective Plan*. Nepal: Ministry of Agriculture, Nepal.

Ministry of Agriculture, Nepal. 2004. *National Agriculture Policy 2004*. Nepal: Ministry of Agriculture, Nepal.

Ministry of Agriculture, Nepal. 2009. *Annual Report 2008/2009*. Nepal: Ministry of Agriculture, Nepal.

Ministry of Agriculture and Cooperatives (MOAC), Nepal. Undated. Available at: http://www.moac.gov.np/ [accessed: May 2011].

United Nations Human Development Index (HDI). 2010. Available at: http://hdr. undp.org/en/statistics/hdi [accessed: 24 May 2011].

World Bank. 2010. *Large-Scale Migration and Remittances in Nepal: Issues, Challenges and Opportunities*. South Asia Region – Report No. 55390-NP (Draft). World Bank.

World Food Programme. 2011. Food Security Monitoring System.

Chapter 6

The Legacy of War: Unexploded Cluster Munitions in Southern Lebanon

Rebecca Roberts[1]

Introduction

As an upper-middle-income country, Lebanon has a well-developed service industry accounting for three-quarters of its gross domestic product (GDP). Approximately 88 per cent of its 4.5 million population lives in urban areas, with the majority concentrated in and around the capital, Beirut (UN FAO 2006: 3–4). These statistics may suggest that agricultural production is unimportant to the Lebanese economy or its population. However, in certain parts of the country, particularly southern Lebanon and the Beqa'a, agriculture plays a major role in the local economy, and contributes around 6 per cent to national GDP (Mottu and Nakhle 2010: 4). The agriculture sector was severely affected by the 2006 war between Hizbollah and Israel, particularly by Israel's heavy use of cluster munitions in southern Lebanon. Cluster munitions are air-delivered or surface-launched munitions containing large numbers of submunitions which are scattered over a wide area, known as the strike area, and designed to explode on impact.

The initial war damage was compounded by the presence of cluster munitions that had failed to explode on impact and, six years after the end of the conflict, continue to deny safe access to land and other natural resources. Israel's use of cluster munitions, and the failure of large numbers to explode on impact, drew worldwide attention to the long-term impact cluster munitions have on civilian lives and activities years after the end of conflict.[2]

1   Interviews were conducted by the author with people living in Lebanon, national and international staff from the Lebanese authorities, United Nations, and national and local NGOs in 2007 and 2011. Thanks are extended to all those who facilitated field visits, particularly the Lebanese Mine Action Centre (LMAC), Mine Action Coordination Centre South Lebanon (MACC-SL), Mines Advisory Group (MAG), Palestinian Women's Humanitarian Organization (PWHO), Sour Community Development Project (SCDP), and individuals affected by cluster munition contamination. Thanks are also due to Clare Collingwood for reviewing this chapter.

2   On 1 August 2001 an international convention to ban the use of cluster munitions entered into international law: Convention on Cluster Munitions (www.clusterconvention. org). By August 2011, 109 states had joined the convention including Lebanon. Israel is not a member of the convention.

The war in 2006 occurred as the agricultural sector was beginning to recover from fifteen years of civil war that had also impeded production and left a legacy of landmines, unexploded cluster munitions and other explosive remnants of war which continue to obstruct agricultural activities 20 years after the end of hostilities. Drawing on research conducted in 2007 and 2011 among the assistance community and local population, this chapter describes the importance of agriculture in Lebanon and the effect the 2006 war and the contamination caused by unexploded cluster munitions have had on agricultural production in the south of the country. This contamination in Lebanon entails long-term problems caused by explosive remnants of war and the time-consuming and expensive process of clearing land. Reducing the threat to the population requires awareness-raising campaigns so that agricultural land can be rehabilitated and cultivated safely.

Agriculture in Lebanon

Although only an estimated 9 per cent of the population of Lebanon are directly employed in agricultural work, it is thought that up to 40 per cent rely on related activities to generate an income (UN HRC 2006: 68). Despite the importance of service industries, in agriculturally based parts of the country up to 90 per cent of the population are reliant on agriculture for their livelihoods, and agriculture represents up to 80 per cent of GDP for the Beqa'a and the south of the country (Darwish, Farajalla and Masri 2009: 630).

Agriculture has the potential to be lucrative, and in the first couple of decades following independence from France in 1945, the future of Lebanon's agricultural sector seemed positive. The climate, fertile soil and good water supply mean that Lebanon has some of the best natural resources in the region. Unfortunately, the lack of investment, technical support and advice, and education means that agriculture has failed to fulfil its economic potential (EIU 2006: 25). Furthermore, the civil war in Lebanon from 1975 to 1990 had a detrimental effect on the agriculture sector, whose share of GDP decreased from 12 per cent before the war to 7 per cent after the war (Darwish, Farajalla and Masri 2009: 630).

The main crops are olives, tobacco, citrus fruits, grapes and vegetables grown on farms ranging from small family-owned plots to commercial enterprises. Breeding of animals for meat and dairy products generally occurs on a small scale, with households rearing a few animals for their own consumption and for the local market. Agricultural products are sold locally and nationally as well as exported. The government regulates the market for some produce such as tobacco, which is purchased by the government at a fixed price for export or to manufacture cigarettes at government-owned factories.

Agriculture is central to the economy of southern Lebanon and represents a significant part of the country's agricultural production. It is estimated that around 30 per cent of agricultural producers in the country are from southern Lebanon, and about a third of those are entirely dependent on agriculture for their livelihoods

(Landmine Action 2008: 28). The Lebanese government believes that all those involved in agriculture in southern Lebanon were directly affected economically by the 2006 war (UN HRC 2006). Before the war, the population of the south was already considered to be among the poorest in Lebanon (UN FAO 2006: vi), so it has few resources to draw on in times of crisis and a limited ability to recover effectively from the impact of war. Assessments post-war predicted that the agricultural sector itself had the potential to recover quickly, but that vulnerable households did not because they would remain burdened by the loss of assets, income and harvest and would become indebted as a means to survive in the short term (UN FAO 2006: vi).

The July War

The 2006 July War fought between Hizbollah and Israel over 34 days from 12 July to 14 August caused deaths and injuries among the civilian population and immediate and long-term damage to the built and natural environment. Although most sectors of the economy suffered heavy losses, industry, tourism and agriculture were the worst hit (Darwish, Farajalla and Masri 2009: 629). Before the war, the World Bank reported that agricultural produce accounted for 20 per cent of total export earnings and it was projected that agriculture would continue to play an increasingly important economic role (World Bank 2006). However, production was so severely disrupted that some farmers in southern Lebanon claim that it will take 20 years to recover from the impact of the war.[3] Agriculture has been particularly affected by the use of cluster munitions because of the large numbers of munitions that failed to explode on impact. The presence of these will continue to pose a threat to civilian life and will deny access to resources for decades. Likewise, the heavy use of cluster munitions by the United States along the Ho Chi Minh Trail during the Vietnam War continues to affect agriculture in Cambodia, Laos and Vietnam four decades later (Handicap International 2007, UNIDIR 2008).

In test conditions, the failure rate of submunitions to explode is a few per cent. This may seem insignificant, but as single warheads are capable of delivering tens or hundreds of submunitions, even a failure rate of 1 per cent can result in millions of unexploded munitions that continue to pose a threat to civilians after the end of a conflict. Failure rates during conflict are often higher than 1 per cent or a few per cent as the military forces using cluster munitions cannot wait for optimum test-like conditions and are operating under the pressures of war, which increase the likelihood of human error. In addition to the psychological pressures of war, other factors determine whether a submunition explodes on impact: the weather, the mode and height of delivery, the type of terrain which if soft or muddy absorbs some of the impact, and the tendency of the parachutes or ribbons that stabilize submunitions in the air to become caught on vegetation or structures thus slowing

---

3   Personal communication, landowner, southern Lebanon, April 2011.

the descent (Hiznay 2006: 22). During conflict, cluster munitions are used in high numbers so even a low failure rate can result in contamination from millions of unexploded submunitions.

During the July War, Israel used a variety of cluster munitions. Following the war, unexploded cluster munitions were found in homes, public buildings such as hospitals and schools, roads, rivers, cultivated land and open countryside. Initially, munitions generally remain visible and on the surface unless they have landed in water or soft ground, have become entangled in vegetation or hidden in structures. Unlike landmines, which are designed to explode when pressure is applied, unexploded cluster munitions do not necessarily explode when touched. This is why untrained individuals without protective equipment can collect large numbers of submunitions without being killed or injured. Clearance by trained or untrained individuals can be rapid while the munitions are still visible. However, with the passing of time and human activity, cluster munitions can become hidden or buried in the ground, posing a greater threat to civilians and slowing down the clearance process. There are numerous incidences, in Lebanon and other countries contaminated by cluster munitions, of individuals harvesting produce from trees, ploughing land or undertaking other agricultural activities being killed or injured because they accidentally strike a munition, or find one in a tree.

One of the warheads used in Lebanon was designed to scatter submunitions over an area of 31,400 square metres (UNEP 2007: 148). Some of the cluster munitions in Lebanon are reported to have a failure rate as high as 40 per cent, and it is estimated that after the conflict, over 1 million unexploded cluster munitions contaminated approximately 34 million square metres of land (NDO 2008: 3). The high failure rate has been attributed to the age of many of the warheads which were over 30 years old and some of the containers failed to deploy any of their submunitions (ICRC 2007: 42). Some experts, however, believe that the air-delivered cluster munitions were deliberately released at too low an altitude in order to create post-conflict contamination, rather than explosions on impact.[4]

When a cluster munition is fired, computer-generated strike data automatically records where it was used. As Israel did not release strike data until May 2009, part of the clearance operation involved identifying strike areas. Consequently, estimates of the number of strikes and number of unexploded munitions have been revised upwards several times as new strikes were found. The data provided by the Israelis added 282 potential strike areas that had not previously been identified (ICBL 2010). By August 2011, 1,277 strike locations had been recorded, contaminating an area of 54,941,844 square metres, of which 133,281 square metres remained to be cleared.[5]

---

4   Personal communication, explosives ordnance disposal expert, December 2010.
5   Email communication, Lebanon Mine Action Centre, 19 August 2011.

Impact of Cluster Munitions on Agriculture

The immediate impact of the July War included almost 2,000 deaths, over 4,000 injuries, the displacement of 1 million people and the destruction of or damage to an estimated 30,000 homes and 900 factories and commercial buildings. The direct financial damage to infrastructure has been estimated at US$1 billion (Darwish, Farajalla and Masri 2009: 629, UNEP 2007: 10, UN FAO 2006: vi).

As a Hizbollah stronghold, the south of the country came under heavy aerial attack which had a direct impact on agriculture. Damage to infrastructure prevented access to markets, although fear of bombing meant that many dared not attempt to transport their harvested goods to market. The sea blockade prevented international trade because goods could not be exported. Landowners and rural labourers fled northwards leaving crops and animals unattended. Crops failed because they could not be irrigated and 80 per cent of poultry was lost because farmers could not feed their birds (Ladki 2006: 3). As a result of the bombing and subsequent contamination, 545 cultivated fields in southern Lebanon were inaccessible (Darwish, Farajalla and Masri 2009: 633). Throughout the country agricultural producers lost up to 85 per cent of their harvest. The loss of the 2006 harvest and the inability to access markets led to many farmers becoming heavily indebted and some were forced to sell their land (Ladki 2006: 3). In the decade preceding the July War, agriculture had achieved an annual growth rate of 2.2 per cent (Word Bank 2006). The war disrupted the agricultural sector's economic growth and has had a 'devastating impact on its productivity' (Darwish, Farajalla and Masri 2009: 630). The cost of direct damage to agriculture has been estimated at US$215 million, with the most commonly cited losses being tobacco, fruits and vegetables (UN HRC 2006: 68).

As stated earlier, the south is one of the poorest areas of Lebanon and one of the most reliant on agriculture for a livelihood (UN FAO 2006: vi). The obvious immediate effect on agriculture was the loss of the 2006 harvest, either because it had been destroyed by the bombing, neglected during a crucial period because farmers had fled, or contaminated with unexploded munitions preventing safe access after the war. The United Nations Environment Programme reported that agricultural land was hardest hit by the cluster munition contamination, which accounted for 62 per cent of all contaminated land. Almost 25 per cent of this land was devoted to olive groves, another 25 per cent to wood and grassland, and a further 15 per cent to olive groves mixed with other crops (UNEP 2007: 155). Seven per cent of grassland used for animal grazing, and the banks and beds of rivers and streams were also contaminated (UN HRC 2006: 68). Tobacco suffered because, although it is harvested only once a year, it must be tended throughout the year to ensure a good crop. The war coincided with the wheat harvest during July and August, so the crop was left to rot in the fields, and olives could not be harvested properly in October and November because unexploded munitions were hanging in the trees.

Removing the Threat

Lebanon's protracted civil war had resulted in large parts of the country being contaminated by landmines and unexploded ordnance. Clearance operations were initiated by the Lebanese Armed Forces (LAF) in the early 1990s. From 2000 there has been significant support from the international community and numerous commercial and humanitarian organizations work alongside LAF to clear explosive remnants of war. Therefore, because Lebanon already had a national demining authority and national and international expertise, as well as specialist equipment for clearing explosive remnants of war, a programme to clear unexploded cluster munitions following the July War could be initiated rapidly. Furthermore, technical experts were able to monitor and record the use of cluster munitions so that some planning for clearance could begin before war had ended. Initially, the mine action authorities estimated that southern Lebanon would be impact-free from cluster munition contamination by the end of 2007. Areas after that time which remained to be cleared would be those that were not densely populated or regularly accessed by the civilian population. Other clearance tasks that had been ongoing in Lebanon were suspended after the war to concentrate on clearing the unexploded cluster munitions.

However, the level of contamination and the time it would take to clear the unexploded cluster munitions were underestimated, and the projected date 'for clearing all unexploded submunitions has been postponed on several occasions since 2006' (ICBL 2010). Funding, which was readily forthcoming from the international community in the aftermath of the war, decreased dramatically by the end of 2009. From 2005 to 2009, national and international funding for clearing explosive remnants of war totalled almost US$174 million. During this period, the most funding was provided in 2006, when contributions reached US$73 million. By 2009, contributions had decreased to US$28 million (ICBL 2010). Understandably, the reduction in funding has led to a reduction in the size of clearance operations. Apart from demonstrating a drop in funding, these figures show the high costs national governments and the international community bear to reduce the threat of explosive remnants of war. In 2008, Landmine Action calculated that the cost over the previous two years and projected costs to complete clearance and undertake related risk-reduction activities, such as raising awareness among the local population of the dangers posed by cluster munitions alone, would total US$120.4 million (Landmine Action 2008: 3).

Of the estimated 1 million people displaced during the war, the Lebanese government and Office of the United Nations High Commissioner for Refugees (UNHCR) figures suggest that between 550,000 and 700,000 had returned home by late August. The rapid return of the population to the south created a greater urgency than had been expected to clear the unexploded cluster munitions. Initial clearance focused on contaminated land – around housing, public buildings and access routes – that posed a direct humanitarian threat. By the beginning of 2007, agricultural land was being cleared. The United Nations Environmental

Programme recommended that agricultural land should be prioritized for clearance to facilitate economic recovery in the south and prevent the use of environmentally unsustainable agricultural practices (UNEP 2007: 156). This might include, for example, overgrazing land or not allowing cultivated land to lie fallow.

In the aftermath of the war, to ensure that clearance operations protected civilians and their livelihoods, the Lebanese mine action authorities identified four different land uses and human activities and prioritized them for clearance accordingly: (1) for the first three months, the focus for clearance would be entrances to houses, roads and highways; (2) in the second three months, the areas around houses and gardens would be cleared; (3) following the clearance of roads and housing areas, agricultural land would be targeted to enable livelihood activities to resume; (4) and finally grazing and hunting land would be cleared. However, by mid-2011 the land around inhabited areas was still being cleared and, despite the clearance of agricultural areas beginning almost immediately after the end of the war, large areas remained to be cleared. The clearance of grazing and hunting land had not begun.[6]

Following advice from the UN Food and Agriculture Organization, the Lebanese Mine Action Centre produced an agricultural calendar to prioritize clearance according to the agricultural cycle so that land could be cleared as it was needed. According to organizations working to clear the land, once it has been cleared it is normally used immediately. This is not always the case in other countries where large-scale clearance programmes are under way to remove explosive remnants of war. The speed at which land is used after clearance in Lebanon shows that there is a pressing need for agricultural land and that the population is anxious to resume its livelihood activities.

The clearance of cluster munitions and other explosive remnants of war is governed by the 2007 Mine Ban Treaty. Although since 2010 there has been a separate treaty banning the use of cluster munitions, in practice the approaches to reduce the threat of all explosive remnants of war are similar and conducted by the same organizations, so the existence of the Cluster Munition Ban has not changed how contamination from unexploded munitions is tackled. All clearance for humanitarian reasons under the terms of the Mine Ban Treaty is undertaken to a very high standard with the aim of removing the threat to the civilian population. Although it is generally accepted that there will be some residual contamination because it is impossible to be sure that all threats have been removed, the process of conducting such thorough clearance is extremely slow. Populations in contaminated areas can become frustrated and often clear land themselves. This happened in Lebanon, particularly in the immediate aftermath of the 2006 war, as farmers wanted to be able to access their land. Wealthy landowners would pay people between US$5 and 7 per submunition cleared; those without the financial resources to pay others to take the risk would try to clear the land themselves. In 2011 there were still people organizing their own clearance. Rather than clear the

---

6   Personal communication, MAG staff, Nabatiyeh, April 2011.

land, some farmers have simply ploughed their fields, which can bury the cluster munitions up to 40 cm deep. This makes them more difficult to find if professional clearance is undertaken in the future.

The relationship between farmers and the clearance organizations can be tense if the farmers feel that the progress is too slow or that their land should be prioritized for clearance. Some farmers refuse professional help to clear their land and sign a waiver to that effect because they do not want their livelihood activities to be interrupted. Whenever clearance does take place, farmers have to be aware that there may be some damage to their crops, for example in orchards and olive groves branches may be cut to enable the clearance to take place. Most farmers accept this as long as it does not mean losing a harvest, and often specifically request that clearance takes place after harvest, thus risking their lives and limbs to work in contaminated fields.

Professional clearance operatives throughout the world face technical and practical challenges and those working in Lebanon are no exception. Without the official strike data from the Israelis, identifying the centres of cluster munition strikes has been difficult and time-consuming. Official emergency clearance in the immediate aftermath of the war, clearance organized by Hizbollah, and spontaneous clearance by the local population increased the difficulty of determining strike areas because the centre of the strike was no longer discernible.[7] Immediately after the conflict, people began clearing rubble from their houses and roads so cluster munitions were moved either deliberately or inadvertently. Clearance organizations had to sift through the rubble to find the cluster munitions.[8] The presence of civilians slows down operations as strict safety procedures govern the work of clearance organizations. Activities cannot proceed if civilians are close by, but ensuring that civilians stay clear of an area during clearance operations is difficult.[9] Those clearing the land also have to maintain a safe distance from each other while working. Therefore, if an area is densely contaminated, it is not always possible to increase the number of people working there to speed up the clearance. Geographical factors also impede clearance activities. Southern Lebanon is mountainous and the steep terrain slows down the rate of clearance.[10] In addition to the dangers posed by the cluster munitions, those working on clearance operations have also suffered from scorpion stings and heatstroke. The heavy protective equipment they wear means that heat and dehydration can be a serious problem in the summer months.

----

7   Telephone interview with Zak Johnson, consultant, Handicap International (France), 23 May 2007.

8   Presentation by Steve Priestly, Director for International Projects, MAG to the Group of Governmental Experts, Convention on Conventional Weapons, Geneva, Switzerland, 20 June 2007.

9   Telephone interview with Andy Glesson, Technical Operations Manager, MAG, 10 May 2007.

10   Author interview by telephone with Anders Wedaa, NPA, 23 May 2007.

Despite the efforts to clear the land, five years after the July War, agriculture production was still not up to pre-war levels. This has had a broad economic impact. One notable change has been to the structure of the labour market which could have significant long-term implications. Lebanese agricultural workers employed by large landowners sought alternative work in the aftermath of the war. Some searched for work in other parts of the country or abroad, others remained in the area and were recruited as soldiers by one of the armed groups, or by clearance organizations and trained to remove cluster munitions and other explosive remnants of war. It is thought to be unlikely that, having found other employment, which may be better paid and more secure than working on the land, these people will return to agricultural labour. Therefore, landowners have been employing migrant workers, particularly Syrians, to fill the employment gap.

Since the end of the civil war there have been large numbers of migrant workers employed in both the agriculture and construction sectors. However, there is concern now among the population of south Lebanon that the number of migrant workers is so high that significant sums of money are leaving the country rather than being redistributed through the local economy. Although Lebanese working abroad send remittances, there is a widespread belief, and therefore concern and fear, that these do not match the amounts leaving the country through the migrant labour force. Furthermore, those leaving the south in search of work tend to be men of working age and this is changing the demographic structure, dividing families and reducing marriage opportunities for women. In a society where the family plays an important role in providing support for individuals, changes in the male:female ratio and age structure of the population threaten the traditional social practices that provide continuity, protection and stability.

In addition to the psychological and social impacts of the cluster munition contamination, there is also a post-war economic impact. In 2008, Landmine Action calculated that current and projected economic costs of lost agricultural production caused by the use of cluster munitions and contamination from unexploded munitions would be between US$22.6 and 26.8 million. Losses in olive production accounted for US$10 to 12 million and citrus fruit another US$4 to 5 million (Landmine Action 2008: 3–4). These losses are significant, particularly in a country that has a public debt of US$47 billion[11] and given that the costs to other economic sectors, and those caused by other types of weapons are excluded from the Landmine Action calculations. Calculating these costs on an individual level suggests that around 3,000 landowners will have lost US$8,000 each in the years following the 2006 conflict (Landmine Action 2008: 2–4).

The number of known victims from the cluster munition contamination is not as high as might be expected. Although the loss of the main breadwinner for an individual family can be a disaster, or the burden of caring for an individual who is unable to contribute to the household income financially draining, the economic

---

11   2011 Lebanon public debt at $47 bln end-2008, Reuters, http://in.reuters.com/article/2009/01/02/lebanon-debt-idINL217217120090102 (accessed 1 August 2011)

impact of deaths and injuries from cluster munitions on the population of south Lebanon as a whole is not considered to be particularly significant. However, the economic perspective should not overshadow the psychological and emotional trauma. Between 2006 and April 2011 there have been 400 victims, of whom 100 have died. Initially it was children who formed the highest number of victims, but now it is adults. It is assumed that this change has occurred because children who were once curious about cluster munitions have been made aware of the danger they pose, but adults are increasingly taking risks and entering contaminated areas to earn a living.

Conclusion

The cluster munition contamination shows how explosive remnants of war have a long-term impact on the civilian population and agricultural production years after the end of hostilities. Lebanon is a small country with contamination from cluster munitions confined to part of it, and therefore, in comparison with the scale of the clearance operations needed to make agricultural land safe in other countries, the efforts needed in Lebanon are relatively small. Nevertheless, in the five years from 2005 to 2009 almost US$174 million have been spent on activities to reduce the threat from explosive remnants of war. During that time, income from agriculture has been lost, with estimates in 2008 suggesting that losses to date and projected losses following the war amount to around US$25 million of revenue (Landmine Action 2008). This is in addition to the direct financial damage to machinery, crops and livestock caused during the conflict. The economic costs can be calculated, but the trauma caused by the death and injury of loved ones, the psychological impact of being unable to generate an income safely and the loss of livelihood activities that may have been in a family for generations, and the damage to traditional social structures and practices are immeasurable. When agriculture plays such a central role among sectors of the population, the inability to resume agricultural activities, not only in the immediate aftermath of war but years after the end of hostilities, has a profound effect on the economic, social and psychological well-being of the civilian population.

References

Darwish, R., Farajalla, N. and Masri, R. 2009. The 2006 war and its inter-temporal economic impact on agriculture in Lebanon. *Disasters*, 33(4), 629–44.

Economist Intelligence Unit (EIU). 2006. *Lebanon: Country Profile 2006*. London: EIU.

Handicap International. 2007. *Circle of Impact: The Fatal Footprint of Cluster Munitions on People and Communities*. Brussels: Handicap International.

Hiznay, M. 2006. Operational and technical aspects of cluster munitions. *Disarmament Forum*, 4. Geneva: UNIDIR.

International Campaign to Ban Landmines (ICBL). 2010. Landmine and Cluster Munition Monitor – Lebanon. Available at: www.the-monitor.org/index.php/ cp/display/region_profiles/theme/399 [accessed: 5 May 2011].

International Committee of the Red Cross (ICRC). 2007. *Expert Meeting on Humanitarian, Military, Technical and Legal Challenges of Cluster Munitions.* Conference report: Montreux, Switzerland, 18–20 April 2007.

Ladki, S.M. 2006. The Summer 2006 Lebanese Food Crisis: A Quality of Life Perspective. Beirut: Lebanese American University (unpublished).

Landmine Action. 2008. *Counting the Cost: The Economic Impact of Cluster Munition Contamination in Lebanon.* London: Landmine Action.

Mottu, E. and Nakhle, N. 2010. Lebanon: real GDP growth analysis. International Monetary Fund Resident Representative Office in Lebanon. Available at: http:// www.imf.org/external/country/LBN/rr/2010/070110.pdf [accessed: 1 August 2011].

National Demining Office (NDO). 2008. *Lebanon Mine Action Program National Demining Office Long-term Plan 2008–2012.* Beirut: National Demining Office.

United Nations Environmental Programme (UNEP). 2007. *Lebanon: Post-conflict Assessment.* Nairobi: UNEP.

United Nations Food and Agriculture Organization (UN FAO). 2006. *Technical Cooperation Programme, Lebanon: Damage and Early Recovery Needs Assessment of Agriculture, Fisheries and Forestry.* Rome: UN FAO.

United Nations Human Rights Council (UN HRC). 2006. *Report of the Commission of Enquiry on Lebanon pursuant to Human Rights Council Resolution S-2/1.* UN HRC.

United Nations Institute for Disarmament Research (UNIDIR). 2008. *The Humanitarian Impact of Cluster Munitions.* Geneva: UNIDIR.

World Bank. 2006. Lebanon at a glance. World Bank.

Chapter 7

Women in Aceh: Conflict and Changing Agricultural Roles

Sylvia Stoll

Introduction

Aceh, located on the most northern part of Indonesia on the island of Sumatra, was severely affected by the 2005 tsunami and the Free Aceh Movement (GAM: Gerakan Aceh Merdeka), which began in the 1970s. The conflict and the tsunami have affected all aspects of Acehnese society; in the aftermath of the tsunami, much attention was given to the recovery of tsunami-affected areas in Aceh. Notwithstanding this attention, the region's three-decade struggle for independence saw a 'drastic rise of poverty, the degradation of public services as well as the increase of human rights violations, with high prevalence of gender based violence' (Brun 2007). The economy of a post-conflict society is always severely damaged and inherently linked to the development of all other structures in society. The economic contribution of women, especially in the informal sector and in agriculture, played a vital role in sustaining Acehnese families during the conflict, as well as in rebuilding socio-economic structures after the conflict.

During a three-week field study in Aceh in July and August 2008, interviews were conducted with 23 women from the districts Aceh Besar, Pidie, Bireuen, Aceh Tengah, Aceh Utara and Aceh Timur to study their economic roles before and after the peace agreement in 2005. Interviews with experts such as employees of local and international NGOs, women activists and academics were conducted for a better understanding of the context and the humanitarian environment in Aceh. Information was also gathered from a focus group discussion with the Center for Community Development and Education (CCDE) group in Pidie, a meeting with a village leader and his family in Aceh Utara, and observing an International Organization for Migration (IOM) progress report meeting in Aceh Utara. Although data was gathered through interviews with a range of actors, documents, publications and participant observation, the majority of the findings here are based on the interviews with conflict-affected women from Aceh.

This chapter focuses on the role of women's livelihoods in Aceh, in particular the changes in these roles before, during and after the conflict, and how specific livelihood strategies are interconnected with the agricultural sector. To set the scene, the chapter discusses the Aceh conflict and the road to the 2005 peace agreement, also referred to as the Memorandum of Understanding. It then describes agriculture in Aceh and women's roles in agriculture in Aceh during and after the conflict. The chapter concludes by summarizing the impact of agricultural changes during and after the conflict and relates this to the future of agriculture in Aceh and the role of women in agricultural production.

The Roots of the Acehnese Conflict

The province of Aceh is populated by about five million inhabitants and covers 12 per cent (57,365 square km) of the island of Sumatra in Indonesia, which is located in South-East Asia bordering Papua New Guinea, East Timor and Malaysia. The Indonesian people consist of multiple ethnic groups and religions, with Islam being the most widely practised.

When Indonesia declared independence in 1945, Aceh became part of the Indonesian republic, but struggled for self-determination. In 1942, Japan invaded and helped the independence activist Sukarno return to declare Indonesian independence. A guerrilla war followed until the Dutch finally recognized Indonesian independence in 1949. Drawing on historical precedent, the Acehnese demanded more self-determination. As a result, in 1950 Aceh was promised special autonomy, but this promise was never actualized.

The frustration of unfulfilled promises was one of the triggers of the Acehnese fighting under Teungku Daud Beureueh against the central government in Jakarta in 1953. Another trigger was the exploitation of Acehnese resources. Aceh's natural gas fields are of great interest to the Indonesian government and the Acehnese feel that their resources were and are being exploited by the central government with little benefit to the local population.

In 1976, Hasan Tiro declared Aceh independent. During the same year, GAM was created, and from that time it fought a bloody conflict with the Indonesian military. From 1989 to 1998 Aceh was a Military Operation Zone. During this time numerous human rights violations took place, including deliberate starvation. With the fall of the dictator Suharto in 1998, peace negotiations were launched between GAM and the new Indonesian government. On 8 November 1999 two million Acehnese demonstrated for a referendum. In 2001 Aceh received permission to implement sharia law. As a result a sharia police was set up and embraced by the Acehnese as a tool for self-determination and 'as proof of the province's autonomy' (Williamson 2006).

After a ceasefire in December 2002, a short period of a free Aceh society was attained. However, the Indonesian army refused to withdraw their troops as agreed and the Acehnese fighters failed to give up their weapons. Six months later, the Indonesian government imposed martial law (Hedman 2005) and deployed more than 40,000 government troops to Aceh. Numerous checkpoints were set up in the province. All foreign humanitarian aid workers and journalists had to leave and the Acehnese were forced to pledge loyalty to the Indonesian government in a ceremony. Most of the victims of this period were civilians, because the Indonesian army attacked both GAM fighters and non-combatants.

It was the December 2004 undersea earthquake off the coast of Sumatra that opened up a new opportunity for peace talks. The tsunami claimed 130,000 death victims, 37,000 missing and 500,000 homeless (BRR and International Partners 2005) in Aceh and again brought attention to the conflict. The Indonesian military was immediately involved in an emergency response in Aceh. The soldiers searched for survivors, buried the dead, and distributed water, food and shelter for the surviving victims. Yet while the soldiers distributed aid, they also continued to hunt for rebels.

Aceh's major losses in the tsunami drew the attention of the international media and NGOs. In light of the severe situation, these actors demanded access to the province. Bakhtiar Abdullah, a spokesman for GAM, acknowledged that it was the disaster that brought the Indonesian government and GAM to the 'negotiating table' (Harvey 2005). The destruction and the shock 'shifted the political dynamics and in effect reinforced what was already true on the popular level in Aceh: a very strong desire for peace' (Renner 2007). The 2005 peace accord was put into practice at the December 2006 elections in Aceh, when Irwandi Yusuf – a former separatist advocate – was elected as governor.

Acehnese Agriculture

Aceh is a region rich in natural resources, yet most people make a living in agriculture and trading. Rice is the staple food, the most important agriculture crop and a major source of employment in rural areas in Indonesia (Winahyu and Acaye 2005). Referring to data from the Ministry of Agriculture, Oxfam reported that Aceh 'had about 359,500 ha of wetland area (*sawah*) and about 1,228,000 ha of dry land area' before the tsunami in 2005. The main crop was rice paddy,[1] 'which accounted for 82 per cent or 295,000 ha of the wetland area' (Winahyu and Acaye 2005).

Other main Acehnese crops are soybeans and peanuts. In addition, a variety of vegetables were grown: 'red onion, chilli, tomatoes, snakebean/longbean, amaranth (bayam), maize, cassava, rockmelon, eggplant, spinach, lettuce, sawi, kangkung (*Ipomea aquatica*), and cucumber', and vegetables 'for waterlogged areas' such as 'kangkung and genger/yellow velvet leaf (*Limnocharis flava*)'. The production of a number of crops like soybeans, chillies and bananas was decreasing until 2004. Coffee (constituting 98 per cent of the exported agricultural products), cocoa, vanilla and patchouli were products that were being exported and produced in small-scale production. Agricultural livelihoods in Aceh normally consisted of 'various combinations of food crops, livestock, estate crops, and fisheries' (Indonesian Agency for Agricultural Research and Development 2008).

---

1   Rice is a semiaquatic crop that Acehnese cultivate in flooded paddies to provide the large quantities of water necessary for irrigation.

Making up 32 per cent of the provincial GDP and almost half of the workforce, agriculture constituted a major part of the Acehnese economy (Nazara and Resosudarmo 2007). Food crops were important both for sustaining families and by contributing to the region's GDP with its rice surplus and export. Most rice production in Aceh traditionally took place on small pieces of land run by peasants. It was common for women to manage these rice fields; they focused on preparing the ground and harvesting, after which the men tended the fields and spread fertilizer. Women were also involved in selling goods in the domestic and public sectors. Although women played a vital role in agricultural livelihoods in Aceh, gender inequalities became apparent in the distribution of landownership. The Rehabilitation and Reconstruction Agency for Aceh–Nias (BRR) 2006 Progress Report states that 'Over 50 per cent of the farmers in Aceh and Nias are women, however women represent a lower percentage of farm and landowners, hence have less decision making power than men' (BRR and Partners 2006).

Women's Agricultural Roles during the Acehnese Conflict

Many peacebuilding activities in Aceh involving women have focused on including them in political processes. However, many of the Acehnese women I interviewed were less concerned about politics and much more interested in feeding their family and surviving economically. Agriculture plays an important role in these endeavours as a large number of post-conflict economies depend heavily on the agricultural sector. 'In many countries where development has recently been curtailed by armed conflict, agriculture was, and is, the primary form of livelihood and the major source of income for a majority of the population' (Sørensen 1998).

During the conflict rice production in Aceh was difficult. Many clashes took place in the rice fields because they served as battlegrounds between the army, which was located in the cities, and GAM, which was hiding in the hinterland. The security issues and the fighting in the field cut the rural population off from the cities and changed their traditional agricultural income activities. The following case studies illustrate the changes in women's agricultural roles that were caused by the conflict.

Case Study 1: Bireuen Family

During my field study I visited an Acehnese school built on a former battlefield as part of a larger reintegration project. The school specifically aimed to employ ex-combatants and reserves a number of free places for children of the victims of the conflict. I met with three female custodians who told me how the conflict had affected them and their families. One of the custodians, a 35-year-old single woman from Bireuen, described how one of her brothers, who was suspected to be a member of GAM, died during the conflict. He was hiding in the forest, where he fell ill and died because of a lack of food. During and before the conflict, the custodian worked in a coffee bean factory and in the rice field. Her mother, however, did not dare to go anywhere because she was afraid. Her other brother was making furniture and is still doing that today. While the custodian and her brothers continued in their previous occupations as rice farmer and furniture maker, the mother discontinued her work as a rice farmer during the conflict and her other brother was hiding from the Indonesian army in the woods.

After the conflict her mother started working in the rice field again, her remaining brother continues his work as a furniture maker and this young woman started to work as a custodian for the school. She is part of an Arisan group, which is a women's group for lending money. She thinks that the most important area to support in peacebuilding is education and wishes she could be a teacher and own a motorcycle.

Case Study 2: North Acehnese Family

During a visit to an IOM event in North Aceh I met with a group of women who were affected by the conflict. One of these women was a 24-year-old mother of one daughter. She was married to a man who had been a member of GAM since 1998. During the conflict the Indonesian army killed her brother-in-law and burnt down her sister's house and shop. The army also took away their animals. This young mother related that she was evacuated to the mountains in 2003, where she stayed for two years. While in the mountains she and her husband harvested fruit, which they sold to buy rice. They also received support from a brother who worked in a restaurant. During this time they ate only once a day and had only rice or sagu to eat. Sagu is the soft part of rumbia wood, which was boiled and used as a food replacement during the war. Now this young mother works in the rice fields that belong to the mosque. In her village every year this land is given to someone else to work on.

Her community earns money by cleaning a fruit together that is called pinang. One shop buys pinang and pays the villagers to come and clean it. This fruit is in season every six months. When the fruit season is over the villagers harvest the leaves of the tree, which they dry and use to wrap a traditional food called dodol. Besides working in the rice fields, the young woman also works on the cleaning of pinang. She wishes that she would buy goats and land so her family could do their own farming.

Both of the women in the case studies undertook agricultural work during the conflict. Some women, however, like the mother in the first case study, did not continue to work in the rice fields because of the clashes that have been taking place since 1976. The experiences in the rice field during the war have left women with trauma due to fighting and sexual violence committed by both GAM and the Indonesian army.[2] During the interviews, several of the women broke into tears as they were reminded of the things they experienced during the war.

---

2 Personal communication with Acehnese women, women activists and NGO representatives in Aceh, July and August 2008.

Today, Acehnese women between 30 and 40 years of age have families and are raising children, but are still suffering from their traumatic experiences during the conflict. Conflicts disrupt economic and social structures, which challenges survival strategies. Women in a conflict environment often carry the burden of being the sole provider for their families in a situation where resources are scarce. In addition, the loss of children and husband can lead to a loss of identity and status (McAskie 1999).

During the conflict Acehnese women were increasingly responsible for providing for their families as many men were either involved in the fighting or unable to work because of it. As a result, Acehnese women played a bigger economic role during the war than they had before. Of the 20 women interviewed, 11 were farming rice during the conflict, three women were engaged in farming crops other than rice, and one worked in livestock farming. Three women reported working on food processing, two on vending or running small businesses (self-employed), two were working in the service sectors, and five mentioned other activities that did not fall within the previous categories. This distribution of income activities demonstrates that agriculture played an important role in women's livelihoods during the conflict, as 18 out of 27 activities were related to agriculture, namely rice farming and other crop farming, livestock farming and food processing.

To deal with the economic and insecurity difficulties they faced, Acehnese women used a number of coping strategies to provide for their families such as finding another job, borrowing money, eating cheaper food, selling something valuable, eating fewer meals per day, sharing income responsibility in the family, and saving energy. The most frequently used coping strategies were to borrow money from a friend (8 out of 18), and to find another job (8 out of 18). During a focus group meeting, a group of women from Pidie described how their economic needs during the conflict forced them to look for additional income. They explained that 'it was difficult to earn a living, because it was not secure and they were afraid of GAM. Nevertheless women had to make a living because the men were not working. They sold vegetables, made cakes (but struggled because nobody bought them), worked in the rice fields, and made chips' (Stoll field notes 2008).

From the data and case studies it is evident that women needed to find additional ways to supplement their livelihoods. During the conflict women's livelihoods focused on agriculture-related activities. Many continued to work in the rice field, and looked for additional or alternative income opportunities. Besides looking for other work, these women also searched for quick ways to access cash by selling jewellery or animals such as goats and chickens. In addition, they changed their diet and 'ate cheaper food like cassava and banana when they did not have enough money' (Stoll field notes 2008). Although there were no changes in the kinds of crops grown and livestock reared, some women used bark they found in the forest to supplement their diets or harvested fruits from the forest instead of working on cultivated land. The young women in the second case study described eating a dish called sagu while living in the mountains during 2003–5. Sagu, made of the soft part of rumbia wood, is boiled and eaten only during times of conflict.

Women's Agricultural Roles after the Acehnese Conflict

The types of income activities that conflict-affected women in Aceh pursued during the conflict differ from those they pursued after the conflict. For example, after the conflict, out of 21 women, five were involved with rice farming and four with other crop farming, four were working on food processing and one on textile production. Three women were vending or running small businesses (self-employed), two were working in service industries and five listed other activities that did not fall within the previous categories. One additional category of income activity – humanitarian or non-governmental organizations (NGO) – was named by a number of women and employed six of the women interviewed.

During the conflict 18 out of 27 activities were related to agriculture; after it only 13 out of 30 income activities were agricultural. Women's farming activities in general have been decreasing since the end of the conflict and agriculture seems to be less important because it is being replaced by other forms of livelihoods. Many more women worked in the rice fields during the conflict to make ends meet and rice continues to be a major agricultural product. However, after the war and especially after the tsunami in Aceh, the numerous NGO livelihoods projects could have influenced this development as they offered new employment opportunities for a number of the interviewed women. The following table compares the income activities during and after the conflict. The first four categories – rice farming and other crop farming, livestock farming, and food processing – are related to agriculture.

Table 7.1   Distribution of women by main income activities, during (n=20) and post-conflict (n=21)

Income activities	During conflict (n=20)	After conflict (n=21)
Rice farming	11	5
Crop farming (other than rice)	3	4
Livestock farming	1	–
Food processing	3	4
Textile production	1	–
Vending/running small businesses	2	3
Working for an organization	–	6
Service sector	2	2
Other	5	5
Total number of activities	27	30

Although the number of women working in agricultural production decreased during the conflict, the aim of a number of NGOs was to strengthen the position of women with projects in agriculture. The 2006 BRR progress report states that 'Gender equity needs to be mainstreamed throughout the rehabilitation and reconstruction process in the agriculture sector to maximize benefits and ensure the overall success of programmes and initiatives to benefit women and families' (BRR and Partners 2006). Often these kinds of activities are also viewed as entry points to underlying societal problems that NGOs aim to address, such as the political role and empowerment of women, the trauma caused during the war or educational objectives and gender issues. In addition, some reconciliation programmes did not directly address agricultural development, but also influenced the decreasing number of women working in agriculture. These programmes provided alternative employment opportunities for a number of women and therefore took them out of the rice fields. The first case study is an example of how a reconstruction programme or project can influence women's agricultural livelihoods in Aceh. The female custodian who had a brother who was suspected to be a member of GAM and died during the war is not working in the rice field and the coffee plantation any more, but is now employed by a school that focuses on reintegration and reconciliation.

Another development that affected agricultural livelihoods after the peace accord was the revival of local financial networks and support systems. This played an important role in the post-conflict rehabilitation of Acehnese agriculture. Forming groups for lending money is a well-known practice in Aceh and often referred to as Arisan groups. Such groups are only formed by women and the Arisan meetings are usually conducted monthly and after Qur'an readings. Every Arisan member contributes the same amount of money at each meeting, although the amount varies depending on the group. According to a system of contribution previously agreed, each week another woman receives the majority of the money that has been collected on that day. These groups are based on trust and could not continue during the conflict because of insecurity. Some of the groups also started businesses together. For example, one business venture financed by a women's group is a local chip-making facility, where a group of women works together to harvest palm tree nuts, roast the nuts, and press them to make a chip that is then packaged and sold. In addition, the second case study provides two more examples of community involvement in agricultural livelihoods. One is the cultivation of rice fields that were managed by the religious community of the village and the second is the harvest of fruit by the community.

On an individual and community level, peace has created new opportunities. Nevertheless, the traces of decades of conflict are apparent in Acehnese households and communities. Because of the war, there are now many more women than men. In addition, ex-combatants who lived in the jungle for ten years find it difficult to do business and pursue livelihoods after the war because they lack the necessary skills and education. All these factors have changed the position of men and women in a household. After the conflict, women's economic contribution to the household increased, but in many cases they again took on a more submissive role in the family. The traditional interpretation of Islamic values and traditional social norms define many types of formal employment as improper for women and therefore affect women's economic opportunities.[3] The need for survival and the absence of many men temporarily shifted these norms and transferred more decision-making power to women. Acehnese women today are more educated and independent than ten or twenty years ago because many of the young women were sent to boarding schools during the conflict for their own security. Increasing educational and employment opportunities have led some women away from agricultural work and are triggers for changes in gender relationships in some cases.

Influence of the Conflict on the Future of Women's Roles in Agriculture in Aceh

The effects of the Acehnese conflict and the current NGO initiatives for empowering women in agriculture have an influence on the future of women's agricultural roles in Aceh. The following list highlights some of the influences of the recent past on women's livelihoods in Aceh and their contribution to the agricultural sector after the conflict:

- Their position as women in an Islamic society
- The gender imbalance created by the conflict left Acehnese communities with more women than men
- The female GAM ex-combatants
- Women's experience of war crimes in the field
- Women's experience of loss of livelihoods during the war
- Some women benefited from attending boarding school during the war
- A smaller number of women are working in the rice fields after the conflict than before
- NGOs are providing alternative employment opportunities to agricultural work
- Women's financial networks were revived after the war.

Throughout this chapter a number of examples have shown how women can work together to improve their livelihoods. Women cooperatives create economic connections between women and therefore their households become interdependent. This provides opportunities to strengthen the community in various sectors such

---

3   For example, during the Pidie focus group meeting, the women talked of a livelihood project undertaken by an international organization where women were taught brick making. Initially the men of the village resisted, as this was seen as an inappropriate profession for women. Eventually the women overcame the men's resistance and participated in the project.

as agriculture, education or new business ventures. A number of organizations such as IOM or the Center for Community Development and Education (CCDE) use community groups to implement their projects. The findings of this study show that women's cooperative groups such as the Arisan groups were and are very important in strengthening local capacities. However, insecurity during the conflict meant that many of these groups were unable to meet. Hence these financial networks play an even bigger role in post-conflict reconstruction and such cooperatives can support post-conflict agricultural activities in Aceh.

At the time of the research in 2008, poverty was still high in rural areas due to the after-effects of the conflict. Women in Aceh experienced three decades of conflict and are known for having taken active roles in the fight against the Indonesian army. During the war when men disappeared or failed to provide income, Acehnese women found ways to cope with economic hardships to provide for their families. When men could not find work, women assumed responsibility to support their families by taking on additional income-generating activities. The strains of the conflict and the loss of Acehnese men left many women with double responsibilities towards their families. While the active fighting ended with the peace agreement in August 2005, women in Aceh continue to fight silent battles against poverty, trauma and loneliness. This affects women working in the rice fields as that is where many of their traumatic experiences occurred. This may be another factor contributing to the decline in the number of women working in rice fields. Although agriculture played a vital role in helping Acehnese families survive, it also exposed women, who had few other options for supporting their families while the men were fighting or living abroad, to insecurity and sexual violence during the conflict.

In addition, the tsunami and the earthquake in Nias in March 2005 had an immense impact on agricultural livelihoods in Aceh. It not only damaged the livelihoods of many Acehnese but also gave the peace process an opportunity to move forward and achieve the Memorandum of Understanding. Also, the tsunami drew the attention of numerous NGOs to Aceh and both the victims of the conflict and the victims of the tsunami are benefiting from projects addressing women's livelihoods and agricultural development. Although in some cases women have resumed their traditional roles and given up certain decision-making powers in their families, the conflict and the after-effects of the tsunami have changed the future of women's roles in agriculture. As more women receive education and therefore access to other opportunities for making a living, they move away from traditional agricultural production. Others benefit from being a member of a cooperative group or a livelihoods project and the financial power this provides for improved and expanded agricultural production. Although changes in Acehnese society and women's roles in agriculture are unrelenting, agriculture continues to play a key role in women's livelihoods. These two case studies on two Acehnese women and their families illustrate the different views. While the woman in the first case study highly values education for peacebuilding and dreams of becoming a teacher, the woman in the second case study wishes to one day own her own land and animals.

References

BRR and International Partners. 2005. *Aceh and Nias One Year After the Tsunami: The Recovery Effort and Way Forward*. [Online]. Available at: http://siteresources.worldbank.org/INTEASTASIAPACIFIC/Resources/1YR_ tsunami_advance_release.pdf [accessed: 19 August 2011].

BRR and Partners. 2006. *Aceh and Nias Two Years After the Tsunami: 2006 Progress Report*. [Online]. Available at: http://reliefweb.int/sites/reliefweb.int/ files/resources/8F7E4BD64414C4C449257252001DDD68-Full_Report.pdf [accessed: 19 August 2011].

Brun, D. 2007. Women's economic empowerment in Aceh and Nias. Oxfam GB, Terms of Reference for Research Consultancy [Online]. Available at: www. siyanda.org/forumdocs/si20070614095802.doc [accessed: 19 August 2011].

Chambers, R. and Conway, G. 1992. *Sustainable Rural Livelihoods: Practical Concepts for the 21st Century*, IDS Discussion Paper 296. Brighton: IDS.

Harvey, R. 2005. Aceh rebuilding gets on track. BBC News [Online, 19 December]. Available at: http://news.bbc.co.uk/1/hi/world/asia-pacific/4542320.stm [accessed 4 June 2008].

Hedman, E. (ed.). 2005. *Aceh Under Martial Law: Conflict, Violence and Displacement*, Refugee Studies Centre Working Paper No. 24. Oxford: Department of International Development, University of Oxford.

Indonesian Agency for Agricultural Research and Development, Indonesia and NSW Department of Primary Industries, Australia. 2008. *A Practical Guide to Restoring Agriculture After a Tsunami*. [Online]. Available at: http://www. dpi.nsw.gov.au/__data/assets/pdf_file/0010/254863/A-practical-guide-to-restoring-agriculture-after-a-tsunami.pdf [accessed: 19 August 2011].

Kim, N. and Conceição, P. 2009. *The Economic Crisis, Violent Conflict, and Human Development*. New York: Office of Development Studies, United Nations Development Programme [Online]. Available at: http://www.undp. org/developmentstudies/docs/hd_conflict_2009.pdf [accessed: 19 August 2011].

McAskie, C. 1999. Gender, humanitarian assistance and conflict resolution. Office for the Coordination of Humanitarian Affairs, United Nations [Online]. Available at: http://www.un.org/womenwatch/daw/csw/Mcaskie.htm [accessed: 19 August 2011].

Nazara, S. and Resosudarmo, B.P. 2007. *Aceh-Nias Reconstruction and Rehabilitation: Progress and Challenges at the End of 2006*. Asian Development Bank Institute [Online]. Available at: http://www.adbi.org/files/ dp70.acehnias.reconstruction.rehabilitation.pdf [accessed: 19 August 2011].

Renner, M. 2007. *Beyond Disasters: Creating Opportunities for Peace.* [Online]. Available at: http://www.wilsoncenter.org/index.cfm?topic_id=1413&fuseaction =topics.event_summary&event_id=244464# [accessed: 6 November 2008].

Sørensen, B. 1998. *Women and Post-Conflict Reconstruction: Issues and Sources,* War-Torn Societies Project, Occasional Paper No. 3. Geneva: United Research Institute for Social Development, Programme for Strategic and International Security Studies.

Williamson, L. 2006. Aceh wary over new Sharia police. BBC News [Online, 8 December]. Available at: http://news.bbc.co.uk/2/hi/asia-pacific/6220256.stm [accessed: 19 August 2011].

Winahyu, R. and Acaye, R. 2005. *Food Aid and the Market in Aceh.* Oxfam [Online]. Available at: http://www.odi.org.uk/resources/libraries/cash-vouchers/oxfam_ cash_aceh.pdf [accessed: 19 August 2011].

PART III
The Recovery of the
Agriculture Sector

Chapter 8

Taking an Agroecological Approach to Recovery: Is It Worth It and Is It Possible?

Julia Wright and Lionel Weerakoon[1]

Introduction

The recovery of agriculture after conflict comprises a complex mix of challenges: those of agricultural development combined with those dealing with the impacts of conflict. Within this, it would be difficult to come across a development intervention nowadays that would not describe its activities as 'sustainable'. Yet just as there are more – and less – sustainable agricultural approaches in stable regions, similarly the types of agriculture that are being promoted in post-conflict situations may also vary widely in terms of their multipurpose effectiveness over time. This chapter addresses the nature of agricultural approaches in post-conflict situations as they relate to sustainability of the natural resource base. Taking as its standpoint that – just as in a non-conflict situation – a more agroecological approach is essential for sustainability; it questions the extent to which current agricultural rehabilitation efforts are taking an agroecological approach and identifies the constraints to this. Two examples are used from post-conflict environments: the resettlement of detainees in northern Sri Lanka, and the rehabilitation of orchards in Afghanistan. Some best practise options are identified.

Why an Agroecological Approach?

The three pillars of sustainable agriculture are generally considered to be economic, social and environmental, with each one being interrelated and requiring attention. In conventional industrial and Green Revolution[2] agriculture, however, the economic dimension has tended to take priority, and environmental or natural resource dimensions have been considered as secondary or non-essential. Yet

---

1   Acknowledgement to Alan Chubb for his involvement in the Afghanistan case study project.

2   Green Revolution agriculture refers to a series of industrial technology developments in the 1960s that involved the development and rolling out of of high-yielding crop hybrids, irrigation infrastructure, and synthetic fertilisers and pesticides to farmers in less industrialized countries.

with the increased understanding of the relationship between food quality and human health, and of the long term detrimental impacts of such an agricultural approach to the natural resource base upon which agricultural livelihoods depend, the more it becomes clear that the environmental or natural resource dimension is essential to maintain if we are to get close to a sustainable system. Agroecology is defined as the science of sustainable agriculture, and is an approach based on an understanding of ecosystems and their inter-relationships. Characteristically it takes a systems and holistic perspective, it enhances natural and life-giving processes, it encourages biodiversity, and it relies upon locally appropriate practices developed through participatory approaches. It is embodied in commonly recognized forms of agriculture such as organic, biodynamic, permaculture and some forms of conservation agriculture.[3] Of all these, it is permaculture that offers up the most comprehensive methodology for the design of sustainable land use, based on ecological and biological principles.

There is a substantial body of evidence on the benefits of a more agroecological farming approach to the natural resource base, as well as to the human resource base. Conventional industrial practices have resulted in vast tracts of degraded land, yield declines, loss of plant and animal species diversity, increase in susceptibility to disease, and other serious side-effects over the medium to long term, and have led to a loss of livelihoods (Hole et al. 2005, Sustain 2003, Oldeman 1999). This is particularly so for marginal lands, where the poor soils cannot sustain monocultures of annual crops, and which are more vulnerable to flood and drought (Hazell and Garrett 2001, McNeely and Sherr 2001). Ecologically based farming practices show themselves to be more successful at supporting a broad and adapted diversity of crop species and varieties, building soil fertility and plant resistance to disease and infection, and maintaining clean water courses (Marriot and Wander 2006, Greene and Kremen 2003). Strengthening the natural resource base also enables farms to better withstand external shocks and stresses, including drought and flood (Ching 2004, Lotter at al. 2003, Holt-Giménez 2002). Nonetheless, more ecologically based production approaches are sidestepped by international development agents owing to their reportedly low yield performance

----

3   Organic agriculture is a holistic production management system that promotes and enhances agroecosystem health, building on biodiversity, biological cycles and soil biological activity. It emphasizes the use of management practices in preference to the use of off-farm inputs, taking into account that regional conditions require locally adapted systems (FAO, 1999). Biodynamic agriculture is similar though its origins stem from a more anthroposophical philosophy. Permaculture uses similar practices in its approach to designing human settlements and agricultural systems that are modelled on the relationships found in natural ecologies. Conservation agriculture, more frequently promoted in the mainstream, aims for sustainable agriculture through the application of particular agricultural techniques such as permanent soil cover and crop rotations, although it does not take as much of a systems approach and may use synthetic chemicals.

and, therefore, their apparent inability to meet global food needs or be appropriate in food insecure situations (IAC 2003).

However, recent studies show ecological farming approaches achieve significant yield increases over both traditional and industrial agriculture, and in particular in resource-poor regions on marginal lands and in tropical and subtropical climates (Delate and Cambardella 2004, Pretty at al. 2002, Parrott and Marsden 2002). Overall, not only is the common uncontextualized focus on yield performance over the short term based on outdated evidence, but it also interferes with achieving food security goals. This focus diverts attention from equally important goals of guaranteeing harvests, increasing community resilience to shocks and stresses, and enabling local availability of a diverse range of quality foods (Wright 2009, Bindraban at al. 1999).

Within the food security debate generally, issues of food quality and diversity have been overlooked. It is not only the natural resource base but also the human resource base that is affected by the farming approach, and this factor is critical in development, humanitarian and post-conflict contexts where human nutrition is a major concern. The practice of industrial agriculture has led to a dramatic decline in the nutrient content of foodstuffs over the last century. This decline is attributed to the unintentional selecting-out of high-nutrient crop varieties when breeding crops for high yield potential, the use of shallow-rooting annual crops that are unable to tap into soil nutrients at deeper levels, and the failure to return a full complement of nutrients to the topsoil. Monocultures also reduce varietal and crop diversity in produce destined for the plate. Whereas a combination of different species results in a more balanced diet than any one species can provide, in particular if using traditional seeds and breeds (Johns at al. 2006, van Rensburg at al. 2004).

Given the above, there is a strong case for mainstreaming an agroecological approach in post-conflict situations. As well as the general benefits of this approach, the conflict may have been caused wholly or in part by environmental scarcity or inequality owing to previous poor management, and achieving stability may involve stabilizing food access and adequacy, both of which an agroecological approach is firmly able to resolve and maintain over the long term. De Koning (2008) directly links natural resources and conflict, through the capture of scarce environmental resources by powerful groups in society and through the inequitable access and distribution of available resources. Diriba (2007) goes further, to highlight the missing link between interventions to end hunger and resource allocation for recovery. Diriba posits that, for communities, it is the inequalities in opportunities and in access to resources that have led to marginalization, and this neglect is often the source of armed conflict. Overall, in post-conflict situations, taking an agroecological approach would encourage localization of resource management, a broader and more diverse range of local opportunities (including higher labour requirements), a more localized food security system, and stabilization of the natural resource base. Its use of traditional approaches would provide some psycho-cultural consistency, and there is also the potential for marketing of higher value organic products.

Two case studies based on the authors' experiences explore the challenges faced when attempting to promote agroecology, and show the degree to which it is possible. The first concerns the promotion of ecological farming to resettled communities in post-conflict areas of Sri Lanka, and the second concerns efforts to develop an export market chain for organic and Fair Trade tree crops from Afghanistan.

Promoting Ecological Agriculture for Resettled Communities in North Sri Lanka

Sri Lanka's population comprises Sinhalese, Moors and Tamils. Tamils are the majority in the Northern Province, of which Jaffna is the capital. The Vaddukodei Declaration of 1977 claimed a separate state for the Tamils, and by 1983 ethnic violence had developed between the Tamils and the Sinhalese majority. Attempts at forging an agreement failed, and in 2007 the government declared war against the Liberation Tigers of Tamil Eelam (LTTE) which ended in defeat for the LTTE. However, the parties most affected by the war were the poorest and marginal sections of the Tamil community living in the conflict-affected areas. Many of these peoples were displaced, including members of the LTTE. Of these, some were detained in army camps and government camps. Initial rehabilitation activities, such as providing immediate assistance for resettlement in temporary zones or areas cleared of land mines were prioritized. Most of the displaceds' houses were damaged or completely destroyed in the war, they had lost their livestock and other belongings, and family members had been killed and injured.

Government resettlement programmes such as 'Uthury Wasanthaya' and 'Nagenahira Navodaya' aimed to improve and construct roads and houses, and public facilities and government offices. However, programmes such as 'Api Wawamu, Rata Nagamu' (we grow and develop the country) and 'Gama Naguma' were slow in achieving their objectives which in turn delayed the next stage of development activities based on the village and community.

Even though they had been given new lands or had their lands returned previously held by them, households lacked the capital to invest in restarting their livelihood activities, the chief of which was agriculture. They did not have funds required to purchase materials necessary to commence agricultural activities. Therefore, organizations such as World Vision initiated programmes to provide cash for work. In addition, displaced communities were receiving World Food Programme aid through a ration system and were also given seeds for agriculture. However, since the seeds were hybrid varieties, communities had to purchase new seeds for every cropping season, incurring expenses that were beyond their capacity to pay for. These hybrid varieties were fertilizer-responsive and involved high production costs. At this time, the government agricultural extension sector was unable to provide adequate support. The private sector, however, intervened by providing more seeds, fertilizers and agrochemicals to farmers and by assuring purchase of their produce. Because these companies were promoting commercial agricultural crops and were less interested in sustainability aspects, households were encouraged to adopt Green Revolution type new agricultural technologies in both irrigated and upland crop cultivations.

These technical changes were markedly different from previous practices. Before the war, farmers were applying traditional practices such as use of cow dung, natural insecticides such as Malabar Nut Tree, and Vogel's Tephrosia that was both a nitrogen fixer and an insecticide. They used their own seeds through a process of selection that guaranteed quality. Mixed cropping was a dominant feature, and they grew perennial crops such as coconut and mango, and semi-perennial fruit crops such as banana. In addition, different kinds of yams, vegetables and green leaves were raised in the home gardens and on farms. Livestock were integrated into their farming systems. These agricultural practices not only helped to maintain soil fertility and soil moisture, but also provided a year round income. Farmers also cultivated commercial crops such as red onion, potato and chilli, for which artificial fertilizer was used. Programmes for re-introducing these traditional technologies, even by NGOs, were not being implemented in the affected areas. One reason for this may have been that NGOs were restricted in their operations owing to security concerns.

Concerned by this situation, representatives of the Sri Lankan NGO Movement for Land and Agricultural Reform (MONLAR), based in Colombo, moved to introduce sustainable agriculture technologies that used natural resources available on the farm or locally and that did not involve a cost. In conjunction with the Mahathma Gandhi Centre and at the request of the Commissioner General of Rehabilitation, MONLAR commenced action to train on sustainable agricultural practices.

Theoretical and practical training was given to LTTE detainees in Kandakaduwa in Polonnaruwa district on sustainable agricultural practices. When it came to the selection of participants, 66 per cent of the 12,500 detainees (including 2,500 women) under the Commissioner General of Rehabilitation expressed their willingness to receive training on agriculture. Out of the 700 detainees at Kandakaduwa camp, 200 were initially trained, and 60 of these were selected to be given extensive theoretical training and practical training to work as trainers. Agricultural activities were also planned for implementation in their villages after they were released from the detention camps. The theoretical training included: current challenges to agriculture, practical training on agroecological techniques for soil fertility (including compost preparation, kitchen waste recycling, liquid manure, natural micro-organisms), mixed cropping, seed conservation, farmers' cooperatives, and transfer of technology from farmer to farmer.

Twenty five widows of Maga Shakthi and Yuga Shakthi Organizations were provided with training on sustainable agriculture at Madampe, Ratmalagara coconut farm, and 1,760 women from Vavuniya, Killinochchi and Mulative districts were given seeds to commence cultivation activities. Most of these women were young, with child dependents, and lack of food was their main problem. They needed this type of agriculture that did not depend on capital for agricultural inputs.

MONLAR, with the involvement of Mahathma Gandhi Centre and Green Movement of Sri Lanka, carried out similar training in 10 villages in Mullaithivu district where less than one third of the population remained compared with pre-war numbers. About 300 participants from these villages received the same training within a period of six days. Around 80 per cent of the participants were women and included disabled people and widows. Agricultural tools such as ploughs, and seeds, were provided in time for the first rains of the main cropping season.

Unsurprisingly, several challenges were encountered when implementing this training programme. Although coordinated activities to resettle the internally displaced and to release detainees and provide them with basic needs were taking place, including baseline data collection, social mobilization, psychosocial programmes, concurrent livelihood restoration activities were not yet being rolled out by the development community. For this reason, agriculture activities in general were held back, particularly in terms of funding support, and thus the crucial rainy seasons were being missed. In addition, the successful implementation of agricultural programmes would require the assistance of security forces and these also had not yet been mobilized. Finally, although the project did not need to supply inputs such as fertilizers and agrochemicals, the returnees still required agricultural implements and seeds to commence agricultural activities, and these inputs were difficult to obtain on a low project budget.

This case study shows that it is possible to train and implement agroecological approaches post-conflict, that these approaches are more appropriate for poor households than industrial approaches, and that detainees are interested to become involved and to reintegrate. However, it also shows that agriculture needs to be considered in the early planning stages of resettlement processes.

Export Market Chain for Organic Tree Crops Kandahar, Afghanistan

Conflict has affected agricultural production in Afghanistan in a number of ways. Insecurity has prevented labourers, input providers and traders from accessing farms at key moments of the productive cycle; an expansion in the urban population due to displacement has affected demand and intensified peri-urban production that is competing with rural producers who are experiencing transport difficulties; changing household composition has reduced family farm labour; the loss of financial assets has limited access to agricultural inputs; displacements have forced some farmers to abandon their farms agricultural outputs and access to land have been distorted by warlords and local militia; formal quality control, regulatory and plant health institutions have ceased to function; changes in the local economy, such as relief food supply, have affected the profitability of staple food production; destruction of common property resources has decreased access by landless farmers; and over exploitation of accessible land areas during the conflict may have long term negative consequences for the natural resource base (Christoplos 2007). In this context, and in an attempt to develop alternative rural livelihoods for the rural poor, a major INGO developed a project to assess the feasibility of exporting organic tree crops from Afghanistan.

During the project's feasibility study, it became clear that development organizations operating out of Kabul were resistant to going out to insecure rural zones, and therefore a large amount of aid funding was being spent on desk studies rather than on practical projects. Consequently, these Kabul-bound offices had little in the way of linkages with international market chains that could catalyse the kind of "trade not aid" export initiative that was strongly requested by Afghan nationals. This particular project, however, undertook a market survey in the UK and identified several fruit importers who were eager to purchase dried fruit produce from Afghanistan – largely because of the 'story' that came with the product. One of these importers offered to come on board in the early stages of project development to support an export market chain. This led to the development of a pilot scale organic conversion and export project for raisins, with a group of 20 farmers from the village of Mandisar, about 3 km outside Kandahar (Wright and Chubb 2006). One of the main long term objectives was to encourage local ownership of the project, and for this to happen, the concept of agroecology or organic agriculture had to be accepted. It was hoped that farmers would be better able to own the organic concept if they saw it as something more culturally meaningful than simply a set of instructions being imposed by a foreign NGO or buyer.

Several strategies were taken to encourage local buy-in of the concept of ecological agriculture. Information and training was provided to both local project staff and the farmers' organization (Mandisar Organic Farmers' Association). This was in the form of comprehensive text books and training packages on organic agriculture, and simple pictorial materials with support to translate these into the local language, Pashtu. During two workshops with farmer representatives, the history of the global sustainable agriculture was related, and emphasis was placed on the resonance of this movement with traditional farming techniques, on its development as a reaction to Green Revolution and industrial agriculture, and on its goals of ensuring food security and environmental resilience over or equal to financial gains. At the same time, a major focus of discussion with farmers was on sustainable and traditional husbandry techniques and their experiences with using agrochemicals and the impacts of these. The link between husbandry practices and perceived food quality was also highlighted. Although many of the changes required in the conversion to certified organic production were of an administrative and logistical nature, a strong experimental husbandry component was also developed to emphasize the roots of ecological agriculture in good husbandry practices. Meanwhile, organic standards were introduced and explained to this new audience as a relatively recent economic coping strategy by the organic movement: a common language for communicating values in the market place. The long distance exporting of organic products was explained in the context of local food systems that import just those foods that could not be grown locally.

Parallel to discussing the export chain, the concept of local markets for quality food produce was raised, and options identified among stakeholders for the subsequent domestic marketing of any produce surplus to project export requirements. In relation to this, farmers were asked for their definition of quality local food. They identified the term 'watani' derived from the Arabic word for 'my homeland' but was commonly used to define domestically produced foods of local breeds and varieties that were generally perceived to be of higher quality than imported, mass produced products. Typical watani produce were domestically produced chickens and eggs, which sold in the marketplace for double the price of imported poultry produce from neighbouring countries such as Pakistan. Descriptive words associated with watani produce were 'tasty', 'luscious', 'sweet', 'big', and 'higher value'. This added weight to the hypothesis that societies have their own local versions of the agroecological concept and of homegrown, organic produce. In this case, traditional, quality produce also commanded a higher price, and poorer households were not excluded from this. Linking the concept of watani with 'organic' enabled project participants to recognize that organic production was not simply an externally-imposed intervention. Further, during discussions, participants pointed out that the agroecological approach to land management resonated with Qu'ranic agriculture, and that the concurrent development of Fair Trade produce struck a chord with Islamic approaches to trading relationships.

Notwithstanding the efforts described above, specific challenges hindered progress of the project and of the full uptake of the agroecological concept by the farmers involved. The project was operated by a foreign NGO, and this fact hindered efforts to hand over the concept of organic as being a local, already-existing feature, or to encourage its local interpretation. At the same time, traditional smallholder agriculture was widespread in Afghanistan, with little historical incidence of high-input Green Revolution approaches or experiences of the negative side-effects of these on the natural resource base. Thus, agrochemicals were not perceived as negative and farmers were not proactively searching for alternatives. There was no national environmental conservation movement, and little environmental thinking or awareness upon which to frame the agroecological approach. The concept of product quality also caused confusion in the project. There was no traditional export market demand for quality produce: farmers involved in the project were historically accustomed to a trading relationship whereby they met the buyers' demands that focused around quantities and prices. For the farmers involved, their highest quality grapes traditionally went to the fresh market, and these grapes were unblemished and large and commanded a higher price than raisins. Low quality grapes traditionally went to the red or black raisin market, these comprising smaller grapes and those that had fallen from the vines, and these were historically dried on the ground and contained dust and stalks. The export project was actually seeking to purchase red/black raisins that were perceived as low quality, and this served to contradict the quality aspect of agroecological production. The project and farmers involved were also influenced by other development initiatives in the country, and several of these contradicted the agroecological message. These developments included the subsidized production and export of industrially-produced grapes, and the development of 'local' produce markets in Kabul that were based on the production and sale of non-traditional crops and varieties that were not local to their environment.

Broader challenges to the project were the security situation that increasingly hindered field visits and deterred international experts – including organic and Fair Trade certifiers – from entering the country; and the short project life (two years) that was insufficient to build up the trust required between project staff, buyers and farmers in order to negotiate realistic business deals (Wright 2009). After two and a half years, the security situation deteriorated to the extent that a decision was made by incoming staff at the contracting INGO to call a halt to the project. Whilst understandable, this decision was not conducive to engendering trust and cooperation with farmer stakeholders in Kandahar. The project could however claim success in that the UK buyer continued to operate in the country and in 2011 was exporting Fair Trade raisins from the more secure Kabul hinterland.

This case study highlights the logistical challenges of operating in a post-conflict environment. It also puts into question the appropriateness of attempting to develop export-focused agricultural operations that impose non-traditional approaches onto smallholder farmers and that may require the build-up of a deep level of trust in order to be successful.

Lessons Learned and Ways Forward

The two case studies show strikingly different approaches to promoting ecological agriculture in post-conflict environments. Both suffer from a lack of financial resources, which indicates that allocation of funds for agricultural projects in post-conflict situations is not immediately forthcoming or is a challenge for NGOs on the ground to access. The Sri Lankan example demonstrates that taking an agroecological approach is worth it, and is possible for resource poor households. Here, the initiative was taken by local organizations in the face of external influences to drive a more industrialized approach to rehabilitation. This example also suggests that agriculture could be brought into the overall resettlement strategy earlier than is currently practiced. In the case of Afghanistan, certified organic production and export was also possible – it eventually took off near to Kabul, but not in the most insecure regions of the country. Here, the main challenge was to support local ownership of the concept of organic farming, because of the externally-driven, market-oriented aspect that required adherence to specific standards and third party inspection by international bodies. A clear difference in the case studies was that Sri Lanka had been using Green Revolution techniques for decades and so farmers had experienced the negative consequences of having to purchase inputs, whereas in Afghanistan this industrial approach had not historically taken hold and therefore the use of conventional inputs was seen to be desirable.

Both examples show that coordinated planning across sectors is crucial, as is a longer planning cycle, which requires a longitudinal planning framework. Methods such as the Resilience Analytical Framework of the FAO (Pingali at al. 2005) have been developed to link the stages of the disaster cycle to improve agricultural resilience. However, these do not fully address the fact that agroecological approaches are relatively less dependent on input provision and more dependent on – and limited by – the degree of ecological knowledge available (Wright 2009). In this sense, building the capacity of the knowledge base – be it through reviving traditional knowledge, introducing ecological knowledge from elsewhere, or supporting sustainable innovation and adaptation for the generation of new knowledge – is crucial. Influencing this is the degree of knowledge held by the intervention organizations themselves, as many comprise personnel who have been trained in conventional agriculture; this includes as much national government agricultural research and extension agents as international. Access to land, and land tenure, is also important for households to invest time and effort in ecological strategies that may show gains only in the longer term. However, it is easier and quicker for interventionists to supply and distribute hard inputs on a large scale than to organize and run training programmes or to go through the process of securing land rights, and even harder for them to acknowledge that they themselves need training. For a transition towards more sustainable approaches, Longley et al. (2007) call for massive capacity building efforts at all levels.

Emerging from these challenges then is the need for a comprehensive agroecological training approach that has fast-acting results over broad areas and that does not require huge funding investments. Of the commonly recognized forms of ecological agriculture, the permaculture system may offer such an approach. Permaculture is a well-documented ecological design system that supports the creation of sustainable, agriculturally productive, non-polluting and healthy settlements. It offers a clear set of principles and practical methods for all agroclimatic regions, and training programmes are short – generally lasting from between two to ten days in duration. In fact it covers not only locally-appropriate agricultural techniques but also relevant aspects of energy, shelter, land tenure and governance (Mollison 1988). Training materials have been developed for refugee situations, and have been applied in refugee camps and settlements implemented by Concern Universal in Malawi, CARE International in Macedonia, and SAFIRE in Zimbabwe. Experiences from these projects have shown that permaculture has several features that make it well suited to refugee situations, including a deliberate spatial design compatible with planned settlements, the development of small scale and intensive production systems, the use of minimal inputs and energy efficient techniques, high labour input, the use of polycultural (as opposed to monocultural) techniques, the production of tangible basic necessities, and the optimal use of water (SAFIRE/UNHCR 2001). Furthermore, evidence indicates that positive results can be achieved rapidly even in the harshest of environments. One permaculture project in Jordan has in one year demonstrated the regeneration of an arid, saline, desert landscape into a productive and resilient farmscape. This transformation was achieved using low cost technology, rainwater harvesting and conservation techniques. After two years, the plant growth success rate was calculated as 90 per cent, and only one-fifth as much irrigation water was required as on surrounding productive areas in the region. Soil salinity levels decreased, as did soil pH levels. Crop yields were similar to those under conventional agriculture on neighbouring farms (ProAct Network 2008). Overall, there is a strong case for researching and developing more appropriate, ecological approaches to post-conflict agricultural rehabilitation. Evidence to date suggests that it is both worth it and possible, and that existing, comprehensive design systems such as permaculture are a good place to start.

References

Bindraban, P., Koning, N. and Essers, S. 1999. *Global Food Security. Initial Proposal for a Wageningen Vision on Food Security.* Netherlands: Wageningen University.

Ching, L.L. 2004. *Organic Outperforms Conventional in Climatic Extremes.* Institute for Science in Society. [Online]. Available at: http://www.i-sis.org.uk/OrganicOutperforms.php [accessed: August 2011].

Christoplos, I. 2007. Narratives of rehabilitation in Afghan agricultural interventions, in *Reconstructing Agriculture in Afghanistan,* edited by A. Pain and J. Sutton. Rugby: Practical Action Publishing.

De Koning, R. 2008. *Resource-Conflict Links in Sierra Leone and the Democratic Republic of the Congo.* SIPRI Insights on Peace and Security No 2008/2. Sweden: Stockholm International Peace Research Unit.

Delate, K. and Cambardella, C.A. 2004. Organic production. Agroecosystem performance during transition to certified organic grain production. *Agronomy Journal,* 96, 1288–98.

Diriba, G. 2007. Reversing food insecurity: linking global commitments to local recovery needs. *Journal of Humanitarian Assistance.* USA: Feinstein International Center.

FAO. 1999. *Organic Agriculture.* Committee on Agriculture 15th Session, 25–29 Jan 1999. Rome: FAO.

Greene, C. and Kremen, A. 2003. *US Organic Farming in 2000–2001: Adoption of Certified Systems.* Economic Research Service, Resource Economics Division, Agriculture Bulletin 780. Washington DC: US Department of Agriculture.

Hazell, P. and Garrett, J.L. 2001. *Reducing Poverty and Protecting the Environment: the Overlooked Potential of Less-favored Lands.* 2020 Brief 39 (June), Washington DC: IFPRI.

Hole, A.G. Perkins, A.J. Wilson, J.D. Alexander, I.H. Grice, P.V. and Evans, A.D. 2005 Does organic farming benefit biodiversity? *Biological Conservation,* 122, 113–30.

Holt-Giménez, E. 2002 Measuring farmers' agroecological resistance after Hurricane Mitch in Nicaragua: a case study in participatory, sustainable land management impact monitoring. *Agriculture, Ecosystems and Environment,* 93, 87–105.

IAC. 2003 *IAC Study on Science and Technology Strategies for Improving Agricultural Productivity and Food Security in Africa*, Progress Report (May) (unpublished), Netherlands: Interacademy Council Study Panel.

Johns, T. Smith, I.F. and Eyzaguirre, P.B. 2006 Agrobiodiversity, nutrition and health, in *Understanding the Links Between Agriculture and Health*, edited by C. Hawkes and M.T. Ruel. Washington DC: IFPRI, 12.

Longley, C.I. Christoplos, T. Slaymaker and S. Meseka 2007. *Rural Recovery in Fragile States: Agricultural support in countries emerging from conflict.* Natural Resource Perspectives 105, February 2007. London: Overseas Development Institute.

Lotter, D.W. Seidel, R. and Liebhart, W. 2003. The performance of organic and conventional cropping systems in an extreme climate year, *American Journal of Alternative Agriculture*, 18(3), 146–54.

Marriot, E.E. and Wander, M.M. 2006. Total and labile soil organic matter in organic and conventional farming systems, *Soil Science Society of America Journal*, 70, 950–59.

McNeely, J.A. and Scherr, S. 2001. *Common Ground, Common Future: How Ecoagriculture Can Help Feed the World and Save Wild Biodiversity*, Report No. 5/01, Gland: IUCN.

Mollison B. 1988. *Permaculture, a Designer's Manual*. Australia: Tagari Publications.

Oldeman, L.R. 1999 Soil degradation: a threat to food security? in *Food Security at Different Scales: Demographic, Biophysical and Socio-economic Considerations*, edited by P.S. Bindraban, H. Van Keulen, A. Kuyvenhoven, R. Rabbinge, and P.W.J. Uithol. Quantitative Approaches in Systems Analysis No. 21, Netherlands: Wageningen University. 105–17.

Parrott, N. and Marsden, T. 2002. *The Real Green Revolution. Organic and Agroecological Farming in the South*. London: Greenpeace Environmental Trust.

Pingali, P. Alinovi, L. and Sutton, J. 2005. Food security in complex emergencies: enhancing food system resilience. *Disasters* 26(9), 4, 283–7.

Pretty, J. 2002. *Agri-Culture. Reconnecting People, Land and Culture*. London: Earthscan Publications.

ProAct Network. 2008. *The Role of Environmental Management and Eco-Engineering in Disaster Risk Reduction and Climate Change Adaptation. Case Study: Jordan Valley Permaculture Project*. [Online]. Available at: http://www.proactnetwork.org.

SAFIRE/UNHCR. 2001. *Permaculture in Refugee Situations, a Handbook for Sustainable Land Management*. Zimbabwe: Southern Alliance for Indigenous Resources (SAFIRE).

Wright, J. 2009. *Sustainable Agriculture and Food Security in an Era of Oil Scarcity, Lessons from Cuba*. London: Earthscan Publications.

Wright, J. 2008. Imposing organic standards or rekindling local values? Encouraging local ownership of the organic concept for raisin exports from

Kandahar, Afghanistan. Paper presented at the 16th *IFOAM World Congress*, Modena, Italy, June 16–20 2008.

Wright, J. And Chubb A. 2006. *Organic Export Feasibility Study, Southern Afghanistan*. Final Project Report, December 2006. Kabul: Mercy Corps.

Chapter 9

Post-apartheid Struggles: Land Rights and Smallholder Agriculture in South Africa

James Bennett

Introduction

Historically, the colonial and apartheid governments of South Africa systematically dispossessed black people of most of their land such that by the time the so-called homelands were consolidated during the 1970s and 80s, the majority of the population occupied just 13 per cent of the country by area (Cousins 2010). The Department of Land Affairs (DLA) estimates that some 16.5 million people, or about 30 per cent of the total population, still reside in these underdeveloped areas (Claassens 2008). Moreover, the state continues to be the *de jure* owner of land in these areas, although in practice the majority of it is held and managed communally by local people (Claassens 2008). Against this political backdrop, land is, unsurprisingly, a highly sensitive issue in South Africa and its ownership and access has been a continuing source of conflict in these communal areas (Ntsebeza and Hall 2007, Cousins and Claassens 2008). This has manifested itself in a variety of ways, ranging from localized occupations of state-owned and private farms (Mokgope 2000) to the contested legitimacy of traditional leaders as custodians of land in communal areas (for instance, Ntsebeza 2005, Oomen 2005).

The importance of land to people in these communal areas stems from the security (as both an economic and physical asset) and the livelihood potential it offers. Although subsistence agriculture now forms a relatively low proportion of overall income for many households in communal areas (Lahiff 2003, Hebinck and Van Averbeke 2007), it still provides an important livelihood stream, which can be drawn upon when required. This is particularly true of livestock production, which takes place on rangelands that are communally held and thereby available to all community members. Livestock grazed on these communal rangelands provide a form of 'living' investment that can be mobilized as cash at relatively short notice if required (Ainslie 2002), and also provide a significant return on investment if all the direct and indirect benefits are considered (Shackleton et al. 2005). Ownership of livestock thus provides an important form of livelihood security, which helps to buffer reductions in other income streams.

The post-apartheid government's attempts at resolving the land and rural livelihoods question have, since 1994, been conducted as part of the Land Reform Programme (LRP) and subsequent developments thereof, which have focused on land restitution, land redistribution and tenure reform (Lahiff 2008). The long-standing government target for land transfer to black ownership (or occupancy) under all aspects of the LRP is 24.9 million hectares by 2014. Although the rate of transfer is now accelerating, overall progress since 1994 has been very slow, with only 4,196,000 hectares of land transferred by March 2007 (DLA 2007). This represents just 5 per cent of the target amount. Moreover, most of these transfers have tended to be on a group basis, often involving hundreds of households as beneficiaries. This has brought considerable challenges in internal group organization, production and the distribution of resultant benefits, and thus limited success in developing smallholder farmers (Lahiff 2008). As a result of these failings, new policy in the form of the Land and Agrarian Reform Project (LARP) has shifted emphasis to projects in rural areas with an explicit sustainable agriculture remit and that focus on supporting emerging communal farmers (DLA 2008).

In communal areas the lack of adequate tenure reform has also been highlighted as a key impediment in securing land rights, reducing conflict and promoting agrarian development (Lahiff 2008, Cousins 2010). Aligned with this has been a lack of progress in reforming and rebuilding the institutions responsible for land access and management. Early attempts at such reform included the Communal Property Association (CPA) Act (1996), which enabled local communities in communal areas to create accountable common property institutions (CPIs) to strengthen property rights and facilitate local resource management. However, where CPIs have been formed they have frequently failed either to adequately uphold their constitutional obligations (Cousins and Hornby 2002) or to manage resources effectively on a collective basis (Bennett et al. 2010). An additional problem has been a lack of external support for CPIs, which no longer appear to be on the political agenda (Lahiff 2008). Rather, there has been a fundamental shift in political emphasis, most clearly embodied in the Communal Land Rights Act (CLARA) of 2004, which provides far more opportunity for traditional leaders to have power over land access and control. Given the often strongly contested legitimacy of chiefs and headmen under apartheid, the Act has generated substantial controversy and has been heavily opposed by many civil society groups (Cousins and Claassens 2008).

In exploring this environment of continuing uncertainty over landownership, access and collective management in communal areas, this chapter draws on two case studies from the former Ciskei homeland of Eastern Cape Province and uses these to highlight cross-cutting issues in the success and failure of current policy aimed at land reform and the concomitant improvement of rural livelihoods in South Africa.

The Case Studies

The subjects of the two case studies that form the core of this analysis have parallel beginnings. Both were former commercial farms that were handed over to black communal farmers during the process of the geographic consolidation of the former Ciskei homeland in the late 1970s and early 1980s. Thus their development, both pre- and post-apartheid, has lessons for current efforts at land reform, particularly tenure reform and the development of effective local institutions for facilitating agricultural development. The two cases are also linked by the involvement of the National Wool Growers' Association (NWGA) at both sites in an attempt to improve wool production, and the relative success of these interventions forms a cross-cutting theme in the analysis.

In documenting the struggles over land and recent attempts to develop collectively managed smallholder agriculture, the work draws heavily on the analysis of rangeland management provided by Bennett et al. (2010), supplemented by original field notes.

Allanwater

The village of Allanwater (Diphala in the local Xhosa language) is situated in Lukhanji Local Municipality, about 15 km south-west of the township of Sada, in the northern part of what was the Ciskei homeland (Figure 9.1).

Background and Historical Development

The core of the current settlement is a former commercial farm of 900 ha that, during the consolidation of Ciskei in the 1970s, was purchased by the government for livestock improvement. However, in 1976 it was occupied by a small group of landless Xhosa migrants and was eventually formalized as a settlement by incorporating it into the local Tribal Authority and appointing a headman and ranger from within the village. The headman presided over a committee of ten people, whose job it was to control all aspects of land access and management at the local level (such as allocation of residential and arable plots and the opening and closing of arable lands as an additional grazing reserve for livestock during the dry season). The headman, in conjunction with the committee, was able to allocate land under Permission to Occupy (PTO),[1] which provided basic security of tenure for landholders.

As a result of this decentralized authority, it is clear even at this stage that the settlement enjoyed a considerable degree of autonomy. For example, it seems that the community decided which grazing camps[2] were to be rested and grazed in a

---

1   PTO was a relatively weak form of land tenure that predominated in communal areas under apartheid, and gave individual rights to a residential and an arable plot as well as communal access to rangeland. Its weakness lay in the fact that individual land rights could be revoked by a traditional leader if 'improper' use was being made of residential sites or inadequate use of arable land (De Wet 1987).

2   Camps are sections of rangeland separated by fencing, which in the case of Allanwater had been established when it was a commercial farm.

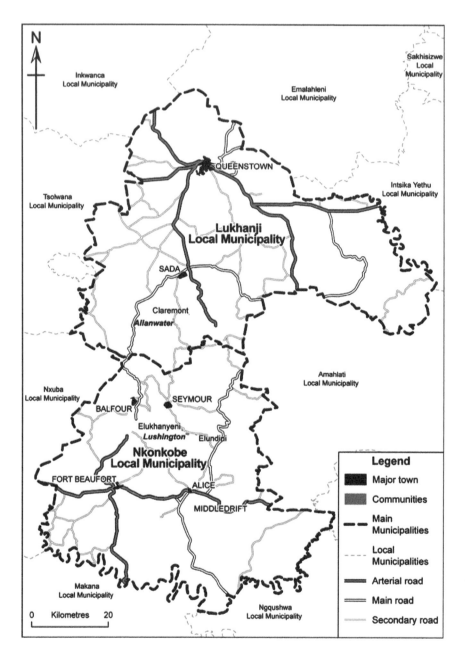

Figure 9.1   Location of case study villages in Eastern Cape Province, South Africa

given season, rather than the District Agricultural Office. The job of the ranger was to enforce these grazing management decisions by ensuring that livestock did not stray onto the camps that were being rested and to check that the fencing was in good enough condition to facilitate this. Another key role of the ranger was to enforce fenced boundaries with neighbouring settlements. This was important to the community as maintenance of forage (the plant material available to animals), both through limiting access by outsiders and resting of the rangeland, was a critical part of ensuring livestock productivity.

With relatively large amounts of land and a small population, agricultural productivity during the apartheid period appears to have been strong. Not only was plenty of rangeland available for grazing but a large amount of arable land was also available for cultivation. Indeed, key informants related that arable land was so abundant initially that individuals could have access to four or five plots each if they had the ability to cultivate them. The importance of livestock to the local community is underlined by the fact that many livestock owners also grew oats and barley as forage for their animals during the dry season (indeed this still continues). This was also supplemented by the grazing of crop residues by livestock once the food crops had been harvested.

Post-apartheid Institutional Change

During the early 1990s, the Ciskei homeland effectively collapsed and traditional leaders were overthrown or, at the very least, marginalized as the institutions of political control in rural areas. With political tensions high, landless people in settlements close to Allanwater took the opportunity to occupy adjacent farms and state land, and radicalized youths destroyed fences and other infrastructure associated with the former apartheid system. These were potentially fraught times for a relatively resource-rich community such as Allanwater. However, the community was able to maintain its integrity and reorganized itself politically in line with the new environment. Thus the former headman and his committee were replaced by a democratically elected Residents' Association (RA), which continues to operate. The RA is constituted by all the adults residing in the village and has a committee of six democratically elected individuals, headed by a chairperson, which administers local issues on behalf of the community. Its main roles include the allocation of residential land and arable plots to those that request them and the general maintenance of law and order. In this respect it is largely inward-facing and appears to have taken over many of the roles of the former headman and his committee. However, it no longer has the power to issue formal title to land as the headman did. It also appears to have a limited role in encouraging externally driven development, and all agricultural development to date has been facilitated by a parallel institution called Vukani Farmers' Association (VFA).

VFA is a legally constituted Communal Property Association (CPA) established in the late 1990s and, like the RA, aims to represent the entire community and encourages every adult to be a member. VFA is run by a committee of six elected members and is effectively an umbrella institution that deals with all aspects of agricultural production in the village. As such it encompasses specialized local producer groups such as the Wool Growers' Association (WGA). VFA is also responsible for all aspects of local agricultural management, including the resting of grazing camps, the maintenance of fencing and the communal dipping of livestock to control external parasites such as ticks. Maintenance of fencing is particularly important, as this not only facilitates grazing management within the community but clearly defines the external boundaries with neighbouring settlements and thereby the extent of local grazing rights. Members of VFA are active in enforcing these boundaries and will expel livestock from outside that gain access. In this respect VFA has taken on the roles of the former ranger, and, in the absence of a formal system of land rights, plays a crucial role in embedding access to and control over natural resource management within the community.

Also of key importance is the external facilitation role that VFA plays. It has strong contacts with both the local and provincial departments of agriculture and has been instrumental in bringing agricultural development to Allanwater. In particular, WGA is very strong in the village and through VFA they have linked in with the National Wool Growers' Association (NWGA) to facilitate a large project aimed at improving wool production. This has involved the subdivision of existing camps to create discrete breeding camps for sheep and the provision of stud rams to help improve the quality of the wool. NWGA have also provided facilities for the shearing, sorting and baling of the wool and have organized transport to take this to market.

Current Agricultural Productivity and Agrarian-based Livelihoods

Almost every household at Allanwater engages with either crop or livestock production and the majority are involved with both (King 2002). All plots of arable land are taken and most are cropped every season. Likewise, around 80 per cent of households at Allanwater own some form of livestock and mean holdings per household in the village in 2002 were 16 cattle, 55 sheep and 18 goats (King 2002), which is considerably more than the regional average in communal areas (Van Averbeke and Bennett 2007). Despite these large holdings, the productivity of the rangeland remains good, with rangeland condition considerably exceeding that of neighbouring communal settlements (Bennett et al. unpublished data). This translates into excellent animal productivity, particularly for sheep. Figures from NWGA for wool production at Allanwater show a total yield of 9,432 kg in 2006 (NWGA unpublished data) or a mean yield per animal of 4 kg. This is comparable to commercial systems in the region and much higher than the 2.3 kg/animal recorded in the communal area of Herschel by Vetter (2003).

This widespread engagement with agriculture provides an important livelihood stream for the community. For most this is supplementary, but some 16 per cent of households at Allanwater are able to make a full-time living out of livestock farming (King 2002). Sheep farming seems to be a critical part of this. Indeed, the total net income from wool sales in 2006 was 226,279 rand (NWGA unpublished data), which averaged across the 43 sheep owners in the village gives a mean income per owner of 5,262 rand. The importance of this cannot be underestimated. Data from the last official household census of 2001 show that mean household income was just 3,473 rand/annum and that more than 50 per cent of households had no formal cash income whatsoever (Statistics South Africa 2001).

That agriculture is able to make such an important and continuing contribution to the livelihoods of local people is, at least in part, a result of the efforts of VFA. Its formation as a CPA, which embraces all members of the community and has an exclusively agricultural remit, has helped to maintain a strong sense of collective ownership of natural resources and a general willingness to engage with agriculture within the village. Whilst this ethos seems to have been a key feature of the Allanwater community since its foundation, VFA was able to galvanize people around this sense of identity at a critical point in the 1990s, when the village was emerging from the institutional vacuum left by the collapse of the Ciskei. Subsequently, VFA has not only continued to manage land-based resources communally but, in its key outward-facing role, has also facilitated external engagement with national programmes such as NWGA, which has brought livelihood benefits to a wide cross-section of the village.

Lushington

Lushington (Cangca in Xhosa) is located in Nkonkobe Local Municipality, about 12 km south-east of the small town of Seymour (Figure 9.1). It consists of several former commercial farms, which previously practised mixed agriculture on a relatively small scale.

Background and Historical Development

The origins of the current village can be traced back to the late 1970s, when the newly established Ciskei government began to buy up commercial farms to consolidate the boundaries of the Ciskei homeland. From this mosaic of released farms three informal settlements emerged, which collectively constitute Lushington. These are Khayelitsha, Elukhanyweni (or simply Eluk) and Elundini. Of these Khayelitsha and Eluk were the first to take root, as former Xhosa farm workers and their families established homesteads on the vacated farms. The origins of Elundini are quite different, with most inhabitants having been forcibly removed from the nearby Tyume Valley area in 1983, to facilitate the building of a dam. Thus, its residents do not share any common ethnic or social ties with the two original settlements. During the mid-1980s, all three settlements were formally surveyed by the Department of Agriculture (DoA), which approved the allocation of residential and arable land.

As the village was formalized in the early 1980s, it was incorporated into the AmaGwali Tribal Authority under Chief Burns Ncamashe and a headman was appointed. He was responsible to the chief and his role was both as intermediary for the articulation of the needs of the village to the Tribal Authority and in settling local disputes such as stock theft. He was supported by a committee who also had a key role in allocating the surveyed areas of residential and arable land, although without formal title. Nevertheless, people still felt that their land rights were secure, as the allocations were made by the committee under the mandate of the government. Arable land was plentiful and cultivation within fenced plots inherited from the former commercial farms was widespread within the village. This was made possible by the farmer support programmes of the Ciskei government, which provided subsidized tractors and other inputs. However, there appears to have been little attempt at centralized control over management of livestock grazing on rangeland despite the inheritance of fenced camps from the commercial farms. Rather, with relatively few livestock in the village, owners simply made unilateral decisions about the management of their animals. Although a ranger was appointed within the village his role appears to have been to check the state of the fences and assess the quality of forage on the camps, rather than control where livestock grazed. The gradual degeneration of the camp fencing during the 1980s further entrenched this laissez-faire approach to grazing management and little attempt seems to have been made by the community to effect any form of repairs.

Post-apartheid Changes in Institutions, Land Management and Agriculture

The effective collapse of the Ciskei as an independent state, in 1990, resulted in widespread rioting in the area by politically motivated youths. Buildings were burnt and the remaining camp fences were cut. This civil unrest continued into 1991, culminating in the headman stepping down from office. During this time of conflict a further breakaway settlement, Ekuphumuleni, was established by disaffected community members on land designated by DoA as arable. This was strongly contested at the time and still causes resentment amongst people from Eluk and Khayelitsha, who see it as an inappropriate use of cultivable land. Nevertheless, the Residents' Association (RA) that was formed in the wake of the political transition of the village is now forced to recognize this as a legitimate settlement.

The current political structure in the village involves a separate committee at each of the four settlements led by an elected chairperson. The RA acts as an umbrella structure for the entire village and is managed by a committee of 12 people, including the four chairpersons from the settlements. It is ostensibly a platform for strategic decision-making across Lushington and for the resolution of issues that cannot be dealt with at each individual settlement. However, in reality it appears to have little power at the local level and, in practice, day-to-day issues such as land allocation and the maintenance of law and order are dealt with by the appropriate local committees. The RA, like the apartheid system before it, also makes no attempt to control grazing management across the village. Indeed, in the absence of most of the perimeter and camp fencing in the village, internal grazing management decisions are now effectively impossible. Not only do livestock from within the village range and graze freely but stock from neighbouring settlements also gain access to Lushington's grazing camps, such that rangeland is now effectively an 'open-access' resource. Thus the community is unable to assert collective rights over the natural resources it should officially 'own'. This not only has implications for the productivity of the rangeland but also makes animals vulnerable to theft or attack by wild animals, if they are left to free-range. In such an environment, only individual decisions over agricultural production are possible and most of these take place at the level of fenced arable plots, inherited from the previous commercial farms. Indeed, fencing is an important determinant of the ability to engage in crop production, as without it crops will frequently be damaged by livestock. Moreover, it also gives field owners the ability to reserve the resultant crop residues exclusively for their own livestock. Given the competition for grazing on the natural rangeland, these residues provide a vital forage reserve for livestock during the dry season and individual rights to them are actively enforced, such that any livestock that stray into them will be quickly removed or even impounded.

Although it has little role in coordinating local land management, the RA at Lushington has been effective in facilitating externally driven agricultural development. As at Allanwater, the NWGA has been active in the village since the late 1990s. This has resulted in the construction of a shearing shed in 1998 and the introduction of improved rams for breeding. A related project instigated by Eastern Cape Department of Agriculture (ECDA) has recently led to the refencing of two small camps to demonstrate the value of resting the range in improving animal production. However, this has had mixed results to date as enforcing this resting at the village level has proved very difficult and the new camp fences have been cut (ECDA 2009). Also of key importance has been a link established with the Massive Food Programme, a government initiative that has been actively supporting crop production in the village since 2003. By providing tractors and necessary field inputs, it has substantially increased people's ability to engage with crop production, following a decline during the 1990s after the Ciskeian farmer support schemes were removed.

Agricultural Production and Agrarian Livelihoods

Current levels of agricultural production and their resultant livelihood impacts are difficult to assess quantitatively due to a paucity of data. Certainly, no figures for crop yields are available either currently or historically to gauge the impact of Massive Food's intervention since 2003. It is, however, clear from key informant reports that the number of households managing to engage with crop production has increased as a result of this support compared with the post-conflict decline suffered during the 1990s. For livestock, the most recent figures for holdings at Lushington show a total of 1,360 cattle, 1,324 sheep and 646 goats (ECDA 2002).

Total wool production in 2006 was 3,395 kg (NWGA unpublished data), giving a mean yield per animal of 2.6 kg, which is fairly typical of output from communal areas but low overall. This low level of wool production (typical commercial outputs in the region are 4.5 kg/animal) is likely to reflect the relatively degraded state of the rangeland at Lushington. Recent range condition assessments by ECDA suggest an increasing amount of unpalatable grasses, herbaceous species and dwarf shrubs in the sward, leading to reduced carrying capacity (ECDA 2009).

Overall Findings and Lessons from the Case Studies

The two case study villages examined here have much in common, both being early examples of land transfers on a group basis and having a shared history in terms of the turbulent political environments that have faced them, pre- and post-apartheid. However, they offer an interesting contrast in the way each community has been able to respond to these changes by adapting its approaches to agricultural production, in the post-apartheid era. Two particular themes emerge in this respect and are analysed below.

Security of Land Rights

One key issue is the security of rights residents are able to exercise over different land resources at each village. Land rights at both villages are now ostensibly informal, with access to residential and arable plots and communal rangeland being granted by local institutions but without provision of formal title. However, this lack of formal title to land does not seem to weaken land rights, providing that local institutions at each community are able to guarantee secure access and control over land. Thus, at both Allanwater and Lushington, the respective RAs act to ensure that households have access to individual plots of arable land and have secure rights of production on these. Consequently, people at both villages have locally secure land rights, which act to incentivize their engagement with crop production.

In contrast, security of rights over communally held land resources such as rangeland differs markedly between the two communities. At Lushington, the RA is unable to guarantee secure collective grazing rights for livestock owners. Rather, in the free-for-all environment that has emerged in the post-apartheid era, livestock are forced to compete ad hoc for the same resources with animals from neighbouring communities and also risk possible theft if they wander too far. In this classic tragedy of the commons scenario, livestock owners thus have access to the resource but no secure grazing rights (Bromley 1989). The situation at Allanwater is quite different. Here responsibility for all aspects of agriculture rests with VFA. As an autonomous CPA, with membership constituted from almost all households within the village, VFA is mandated to take collective management decisions on behalf of the community, which are respected and largely upheld by the residents. It also has a crucial role in enforcing resource boundaries with neighbouring communities. Thus, members of VFA have secure rights to grazing as part of a well-defined user group, which has clear resource boundaries and a common set of rules for resource use; all of which are established prerequisites for an effective common property regime (Bromley 1989).

Thus, it is apparent from these examples that land rights can be secure even in the absence of formal title. However, under these circumstances the extent of this security depends almost entirely on the ability of local institutions to enforce rights of access and control over land at either an individual or community level.

Engagement with Sustainable Agriculture

Another important feature of the development of each village in the post-apartheid environment has been its ability to promote sustainable agriculture as a rural livelihood stream. As part of this, both villages have been able to engage with externally driven agricultural programmes. A case in point has been the involvement of Massive Food at Lushington, which has increased the ability of local households to undertake arable agriculture. Nevertheless, whilst this has undoubtedly provided an important additional livelihood stream for many households in the short term, its viability in the longer term remains questionable. The equipment and input costs are heavily subsidized by the programme and it is unlikely that current crop production levels would continue without this. Thus, there is the inherent danger of a collapse in production, as occurred following the demise of the farmer support programmes in the former Ciskei, should these subsidies be reduced or removed. Interestingly, arable agriculture at Allanwater receives no such support but crop production is, arguably, even more widespread as almost every household has access to at least one field and most are cultivated every year. They are able to do this because they have maintained an independent ability to farm, making use of animal traction and local inputs, and, in this sense, arable agriculture at Allanwater is sustainable.

The other contrast between the villages has been in their ability to respond to recent initiatives aimed at improving livestock production from rangeland. Since the late 1990s, both communities have been involved with projects, instigated through NWGA and ECDA, to try and improve wool production. At Lushington, despite the construction of local infrastructure and the provision of improved breeds of rams, these efforts have had limited impact due mainly to the inability to improve the quality and quantity of available forage. This has proved impossible in an environment where the community is not only unable to control where its own stock graze, but also cannot exclude animals from neighbouring settlements. There has also been a lack of collective support from the community for some projects, such that attempts by ECDA to rest areas of range by reintroducing fenced camps have been persistently compromised by incidences of fence cutting. Forage remains poor and the rangeland shows clear signs of localized and continuing degradation, which brings into question the sustainability of livestock farming as a whole in the village in the longer term. In contrast, wool production at Allanwater has been improved by the intervention of NGWA and is providing a significant source of income for sheep owners. This is primarily due to VFA's ability to manage grazing resources to ensure a continuous supply of natural forage, augmented by the willingness and ability of owners to grow forage crops for animals during the dry season. Providing resources continue to be judiciously managed and livestock numbers do not outgrow the available grazing, it is likely that this will continue to be a sustainable and productive enterprise for the foreseeable future.

These case studies demonstrate that in communal areas secure land rights are an important prerequisite for agriculture to occur successfully and underline the critical importance of effective local institutions, particularly CPIs, both in securing these rights for local people, especially to commons resources, and in brokering external support from government projects. However, an important point that also emerges from this is that the existence of such institutions is very much dependent on the local social and political environment. Where communities are fragmented through ethnic and political divisions, as is the case at Lushington, such institutions are highly unlikely to emerge, or to be sustainable in cases where they do. Rather, production is likely to devolve to an individual level because of a lack of collective identity. Only where there is a supportive environment, which can foster a shared desire to engage with agriculture on a community basis, as exemplified by Allanwater, are institutions such as CPIs viable and capable of providing the secure land rights necessary to support collective agriculture.

Conclusions

These findings have important implications for current land reform efforts in South Africa. First, the fact that local land rights can be secure without formal title, as long as they are guaranteed by local civil society institutions, suggests that the current political approach to land reform may be misguided. Recent attempts to formalize land rights in communal areas through legislation such as the Communal Land Rights Act (CLARA) have advocated formal land survey with issue of title. This may be unnecessarily expensive and unwieldy. In areas such as the former Ciskei, simply formalizing existing arrangements may suffice, particularly in securing rights to arable plots. Securing community rights over communal grazing resources may be less straightforward, but where strong CPIs are in place it is clear that this is still possible on a civil society basis.

This suggests that the government's abandonment of support for Communal Property Associations (CPA) on the basis of their general failure to demonstrate livelihood improvements may be premature. Lessons must be learned from failed pre- and post-apartheid land transfers to encourage a more rigorous case-by-case evaluation of where the CPA approach might be appropriate and where social and political divisions within communities may constrain this. Given adequate legislative and community-level support, the evidence suggests that such institutions can be viable and, at the very least, should continue to be supported where they are already producing livelihood benefits. Importantly, CPAs also offer a more democratic alternative to traditional institutions such as chiefs and headmen, whose revival has featured prominently in recent attempts at land reform such as CLARA.

Finally, the findings also offer hope for the current LARP approach to land reform adopted by the South African government. By working to actively support communities such as Allanwater, which have demonstrated a willingness and ability to embrace the agricultural opportunities presented to them, it may be possible to develop a class of commercially oriented farmers in communal areas. Recent efforts by NWGA to assist communal sheep farmers in Eastern Cape Province show that government-driven agricultural interventions can be successful, given adequate support to appropriate communities. However, it is important that the government recognizes that this success depends on identifying those communities where there is a collective willingness to engage with such initiatives and where appropriate institutions for collective natural resource management either already exist or can be readily developed. Without effective CPIs that are able to command such widespread community support, such interventions will fail or provide livelihood benefits to only a few. Identifying and promulgating examples of best practice in CPI organization and development will therefore be imperative if efforts by organizations such as NWGA are to find wider success.

References

Ainslie, A. (ed.). 2002. *Cattle Ownership and Production in the Communal Areas of the Eastern Cape, South Africa.* Research Report Number 10. Programme for Land and Agrarian Studies, University of the Western Cape, South Africa.

Bennett, J., Ainslie, A. and Davis, J. 2010. Fenced in: common property struggles in the management of communal rangelands in central Eastern Cape Province, South Africa. *Land Use Policy,* 27(2), 340–50.

Bromley, D.W. 1989. Property relations and economic development: the other land reform. *World Development,* 17(6), 867–77.

Claassens, A. 2008. Power, accountability and apartheid borders: the impact of recent laws on struggles over land rights, in *Land, Power and Custom: Controversies Generated by South Africa's Communal Land Rights Act,* edited by B. Cousins and A. Claassens. Cape Town: UCT Press, 262–92.

Cousins, B. 2010. The politics of communal tenure reform: a South Africa case study, in *The Struggle over Land in Africa: Conflicts, Politics and Change,* edited by W. Anseeuw and C. Alden. Cape Town: HSRC, 55–70.

Cousins, B. and Claassens, A. (eds). 2008. *Land, Power and Custom: Controversies Generated by South Africa's Communal Land Rights Act.* Cape Town: UCT Press.

Cousins, T. and Hornby, D. 2002. *Leaping the Fissures: Bridging the Gap Between Paper and Real Practice in Setting Up Common Property Institutions in Land*

Reform in South Africa. Land Reform and Agrarian Change in Southern Africa paper no. 19. Programme for Land and Agrarian Studies (PLAAS), University of the Western Cape, South Africa.

De Wet, C.J. 1987. Land tenure and rural development: some issues relating to the Ciskei/Transkei region. *Development Southern Africa*, 4(3), 459–78.

DLA, 2007. *Annual Report, 2006–2007*. Pretoria: Department of Land Affairs.

DLA, 2008. *The Land and Agrarian Reform Project (LARP): The Concept Document*. Pretoria: Department of Land Affairs.

ECDA, 2002. *Livestock Census Data*. Dohne, Stutterheim: Eastern Cape Department of Agriculture.

ECDA, 2009. *Veld Management and Resting of Communal Rangeland in the Eastern Cape, Annual Report 2008–09*. Project no. 5411/41/1/9. Dohne, Stutterheim: Eastern Cape Department of Agriculture.

Hebinck, P. and Van Averbeke, W. 2007. Rural transformation in the Eastern Cape, in *Livelihoods and Landscapes: The People of Guquka and Koloni and Their Resources*, edited by P. Hebinck and P.C. Lent. Leiden: Brill, 33–66.

King, B.R. 2002. The Establishment of an Effective Farming System for the Allan Waters Communal Area in the Eastern Cape Province. Unpublished MTech thesis. Department of Agriculture, Port Elizabeth Technikon.

Lahiff, E. 2003. *Land Reform and Sustainable Livelihoods in South Africa's Eastern Cape Province*. Sustainable Livelihoods in Southern Africa, Research Paper 9. Programme for Land and Agrarian Studies (PLAAS), School of Government, University of the Western Cape.

Lahiff, E. 2008. *Land Reform in South Africa: A Status Report 2008*. Programme for Land and Agrarian Studies (PLAAS) research report no. 38, School of Government, University of the Western Cape.

Mokgope, K. 2000. *Land Reform, Sustainable Rural Livelihoods and Gender Relations: A Case Study of Gallawater A Farm*. Programme for Land and Agrarian Studies (PLAAS) research report no. 5, volume 1. PLAAS, School of Government, University of the Western Cape.

Ntsebeza, L. 2005. *Democracy Compromised: Chiefs and the Politics of Land in South Africa*. Leiden: Brill.

Ntsebeza, L. and Hall, R. (eds). 2007. *The Land Question in South Africa: The Challenge of Transformation and Redistribution*. Cape Town: HSRC Press.

Oomen, B. 2005. *Chiefs in South Africa: Law, Power and Culture in the Post-Apartheid Era*. Oxford: James Currey.

Shackleton, C.M., Shackleton, S.E., Netshiluvhi, T.R. and Mathabela, F.R. 2005. The contribution and direct-use value of livestock to rural livelihoods in the Sand River catchment, South Africa. *African Journal of Range and Forage Science*, 22(2), 127–40.

Statistics South Africa. 2001. *2001 Census Data*. Pretoria: Department of Statistics.

Van Averbeke, W. and Bennett, J. 2007. Agro-ecology and smallholder farming in the central Eastern Cape, in *Livelihoods and Landscapes: The People of*

Guquka and Koloni and their Resources, edited by P. Hebinck and P.C. Lent. Leiden: Brill, 67–90.

Vetter, S. 2003. What Are the Costs of Land Degradation to Communal Livestock Farmers in South Africa: The Case of the Herschel District, Eastern Cape? Unpublished DPhil thesis. Department of Botany, University of Cape Town.

Chapter 10

Cambodia: The Challenge of Adding Value to Agriculture after Conflict

Sigfrido Burgos Cáceres and Sophal Ear[1]

Abstract

Conflict can trap a country in poverty and economic stagnation. This chapter presents a case study of agriculture's role in post-conflict Cambodia. Opportunities exist for improving rural livelihoods in Cambodia through the livestock sector. Even more promising is the rice sector, but only if rice milling (processing) can emerge and replace large unofficial exports of husked rice to neighbouring countries (primarily Vietnam). Value addition has not been Cambodia's forte. Agriculture's primacy is undeniable, given that it is the only sector capable of absorbing the 300,000 entrants joining the labour force each year. Agriculture is paramount in the Cambodian economy, accounting for 40 per cent of GDP. To Cambodians, agriculture is not only a way of life but a tool to ensure survival. Unfortunately, investments (for instance in markets, transport, processing) have been too few and far between to unleash the potential of the agriculture sector owing to weak state capacity and poor governance.

Introduction

Throughout its modern history, Cambodia has been plagued with conflict.[2] For instance, from 1955 to 2009, Cambodians suffered perennial *coups d'état*, incursions, wars, skirmishes and a genocide under the Khmer Rouge.[3] During and after these conflicts, agriculture suffered owing to arable land infested with landmines and unexploded ordnances (UXOs), corruption, mass deaths, land

---

1   Disclaimer: The views expressed in this information product are those of the authors and do not necessarily reflect or represent the views of the United Nations, the Food and Agriculture Organization, the US Navy, or the US Government.

2   By 'conflict' we refer to battles, combats, conflagrations, fights, shootings, skirmishes and wars.

3   The Khmer Rouge, literally Red Khmer, was a Maoist rebel group that took control of Cambodia on 17 April 1975 and was deposed when Vietnam invaded Cambodia and took over Phnom Penh on 7 January 1979.

confiscation and redistributions, trade restrictions as part of internationally imposed sanctions, damaged infrastructure and domestic market disruptions, among other man-made calamities (Chandler 2007, Colleta and Cullen 2000).

As demonstrated by Collier et al. (2003), armed conflict can trap a country in poverty and economic stagnation. This outcome is not surprising. If one considers land, people, money and markets as the four fundamental pillars of agriculture, all of these are disrupted during and after conflict. Furthermore, the larger sociological implications of conflict include societal reordering that affects economic activities by redistribution of manpower from productive to unproductive outputs (Modell and Haggerty 1991).

International economists, using data sets from 58 low-income countries, estimate that the cost of a typical civil war to the country and its neighbours is approximately US$65 billion (Collier 2008: 32). As if it were not enough, the spillover effects of conflicts persist in low-income countries through, for example, pervasive civil unrest, extortion, illicit trade of goods, robberies, labour exploitation, pillage, land grabbing, widespread corruption, illicit resource extraction and other criminal activities.

More specifically, one could argue that disruptions to the primary sector of the economy[4] prove detrimental to agriculture-dependent countries given that its contribution to GDP can represent anywhere between 30 and 80 per cent. This is the case with Cambodia, where agriculture is paramount in the economy at currently 40 per cent. The numbers speak for themselves: agriculture as a percentage of GDP[5] was on average 55 per cent in the 1980s, 47 per cent in the 1990s, and 40 per cent in the first decade of this century (IMF 2010, World Bank 2010).

The slow yet steady decline of agriculture's contribution to GDP over the last thirty years is in part explained by Cambodia's government – now a constitutional monarchy – implementation during the mid-1990s of a series of neoliberal market policies. The overall macroeconomic performance derived from these interventions has been positive for GDP. For example, support to the agricultural sector improved production of some commodities, especially rice; whereas support to the industrial and service sectors opened up the country to the world not only as a location for foreign-owned manufacture of garments but also to tourists wishing to explore Cambodia's natural and cultural treasures such as the temples of Angkor Wat.

---

4 In short, this economic sector is involved in transforming natural resources into primary products that are considered raw materials for other industries. Major enterprises include animal and crop agriculture, agribusiness, fishing, forestry and mining.

5 The GDP calculation method, principally established in the 1930s, is precise in its ability to account for all goods and services produced, but woefully imprecise in accounting for human and natural resources. On the other hand, the genuine progress indicator (GPI) measures the sustainability of multifactor national income and the socio-economic well-being of a nation. To put their differences into context, from 1950 to 2010, global GPI has grown at a much lower rate than global GDP (Gore 2009: 324).

While this rebalancing of economic sectors has proved largely accretive to the country, it is worth noting that 81 per cent of the population of 14 million is still classified as rural and, as result, agricultural activities cannot be divorced from economic development. In this context, agriculture's primacy is undeniable, especially since it is currently the only economic sector capable of absorbing the 300,000 entrants joining the labour force each year (Ear 2005). In short, to Cambodians, agriculture is not only a way of life but a tool to ensure survival (Burgos et al. 2008).

So far, little has been said about how conflicts have affected agricultural production in Cambodia. The reason for this is that no comprehensive analysis has been conducted in this regard.[6] To give an idea of the economic effects of conflicts on the agricultural sector, we provide a rough estimate: if conflict for any given year in the 1990s resulted in a 1 per cent drop in total GDP from an average of US$30 billion, this means that US$300 million was lost in the Cambodian economy due to conflict.

Clearly, this figure overemphasizes the economic cost of conflict while it understates the social and cultural value losses to what could be considered a cohesive agrarian society with tightly woven communal safety nets (Marks 2000). The precision of calculations is not the point. The point to put across is that conflict can be interpreted through the duality of bad and good: bad because it invites death, injury, insecurity, poverty, stagnation and suffering, and good because it also welcomes aid, assistance, cooperation, opportunities, reassessments and resilience.

A closer look at Cambodia's trajectory after gaining peace in 1991 shows that state capacity building, openness to investments and structural reorganization, among other factors, brought back robust economic growth to the country. Furthermore, the vibrancy of a few dynamic sectors, notably garments and tourism, each contributing around 15 per cent to GDP, has provided labour opportunities to a large number of unskilled rural workers as well as primary and secondary infrastructure developments into predetermined industrial production zones (Ear 2009).

Although these demand-side endeavours command praise, ever-present risks threaten their continued success: global economic gyrations that modulate consumption levels in high-income countries suggest that garments are prone to cyclical (and sometimes protracted) slowdowns; while emerging infectious diseases such as Nipah in pigs, Severe Acute Respiratory Syndrome (SARS), H5N1 Avian Influenza and Pandemic H1N1 Influenza cause mortality and morbidity,[7] but also scare off visitors who contribute significantly to the tourism sector (Cáceres and Otte 2009).

Another shining star in the primary sector has been growth in resource-extractive activities: the extraction of gold, timber, precious stones, oil and gas. The

---

6   The economists Anke Hoeffler and Paul Collier estimate the typical cost of civil wars at US$60–70 billion.

7   For a short explanation of the multidimensional linkages between animal diseases and international affairs, see the article by Burgos and Otte (2011).

vast opportunities for profit have fuelled conflicts and enabled graft and greed to become widespread (Burgos and Ear 2010). To be sure, while agriculture, drilling and mining require not just a measure of security,[8] a dynamic, entrepreneurial and formal private sector within Cambodia is pivotal to ensure that windfalls can trickle down to its society in the form of rising household incomes, increasing domestic consumption, growth of construction and real estate, and renewed opportunities for regional and international trade (Ear 2005).

Through a literature review that covers an array of books, scholarly articles and institutional working papers, coupled with the authors' experience with the country, this chapter focuses on the challenge of adding value to agriculture after conflict in Cambodia. It starts by presenting a country background along with a brief explanation of country-specific agriculture and conflict history; it then focuses on animals and crops, followed by an elucidation of the difficulties of working in agriculture in post-conflict environments in the tropics. Lastly, we reflect on the vicissitudes of Cambodia's experience and the lessons learned from them.

Country Background: Geography, Agriculture and Conflicts

The Kingdom of Cambodia, more commonly referred to as Cambodia, is a South-East Asian country that borders Thailand to the west and north-west, Laos to the north-east, Vietnam to the east, and the Gulf of Thailand to the south-west. The capital of Cambodia is Phnom Penh, and the official language and script is Khmer. With an area of 181,035 square kilometres, the country is crossed by the Mekong River, with a lake in the middle called Tonlé Sap. Agriculture has long been the most important economic sector, with rice being the principal crop. Cambodia's primary export is garments, although it has attempted to make inroads with livestock, and may someday export natural gas and crude oil, market conditions permitting. Like many other agriculture-oriented nations, Cambodia's agricultural exports are rice, fish, timber and rubber with little added value. This means that after harvest, almost nothing is done to these items in terms of additional processing or packaging so as to give them a competitive edge in the marketplace. As a norm, products with added value command higher prices.

Because the Cambodian territory is nestled in among Laos, Thailand and Vietnam, in the fifteenth and sixteenth centuries it experienced a series of wars with neighbouring kingdoms. Later, from 1841 to 1845, the Siamese–Vietnamese War ended with an agreement to place Cambodia under joint suzerainty. By 1863, King Norodom, who is credited for saving the country from disappearing altogether, sought the protection of France from the Thai and Vietnamese, after tensions grew between them. After 90 years of colonialism, on 9 November 1953

---

8   The perceived high risks of investments in volatile, conflict-prone, low income economies deter significant inflows of private capital needed to pay for equipment, inputs, and value-adding processes (Collier and Pattillo 2000).

Cambodia gained independence from France. After a series of internal power struggles, American B-52s bombed Cambodia to the tune of five Hiroshimas and Nagasakis in an effort to disrupt the Viet Cong and Khmer Rouge between 1969 and 1973. The darkest period of Cambodia was 1975 to 1979 when the Khmer Rouge exercised its communist-led authoritarian power that resulted in the death of nearly two million people or a quarter of the population.[9] The killing of the productive labour force plunged the country into agricultural decline.

Exploiting Animals or Crops? A Question of Convenience and Expediency

International and national interventions aimed at strengthening the livestock sector are relevant for reducing hunger and alleviating poverty as thousands of rural households rely heavily on livestock production for sustaining their livelihoods. Regrettably, evidence from the developing world suggests that the true potential of the livestock sector has been underexploited because animal production has been regarded only as an appendage to agriculture and not as an integral contributor to agricultural GDP. In fact, both development practitioners and national decision-makers have long given utmost priority to staple crops[10] over high-value agricultural outputs such as raw or processed animal-source foods, fruits and vegetables (FAO 2010: vii). Additionally, very little attention has been paid to incentives, agricultural policies and consumption markets that are critically essential for the successful development of the livestock sector (ILRI 1995). In Cambodia, many opportunities exist for improving rural livelihoods through the livestock sector, be it through the strategic exploitation of cattle (beef and dairy), pigs or poultry.[11] These are also agricultural products that can gain from value-adding.

To disregard high-value agricultural activities in countries where agriculture is the main engine of growth, purveyor of food security and insurance against riots and widespread malnutrition is not only myopic but downright irresponsible. Coincidentally, as Bill McKibben (2007: 199) rightly asserts, governments and aid agencies do not pay much attention to rural initiatives and local agricultural institutions because, in part, they get in the way of trends to push for manufacturing and industrial production. This could not be truer in Cambodia. But it is not as easy as people think. Some fundamental prerequisites need to be in place for agricultural subsectors to succeed. For instance, a workshop report from the Center for Stabilization and Reconstruction Studies at the US Naval Postgraduate School notes that in post-conflict societies, like Cambodia, there are four main

---

9   See various books by David Chandler, Phillip Short, Joel Brinkley and John Tully.

10   Staple crops are cereals such as sorghum, wheat, barley, rye, maize, rice, or starchy root vegetables such as potatoes, yams and cassava.

11   In value terms, poultry is still the smallest of these three livestock activities. It is considered small change for backyard farmers, whereas the sale of a cow for a wedding or a funeral is common.

investments in agriculture: agricultural research, extension services, credit provisions and infrastructure development (CSRS 2010).

Before going any further, a look at the numbers of animals present can set the stage for the ensuing narrative: in 2005, livestock numbers in Cambodia were estimated at 3 million cattle, 0.65 million buffaloes, 2.5 million pigs and 22 million poultry (FAO 2007: 113–25). Three years later, by 2008, modest changes had occurred to livestock numbers: these were estimated at 3.5 million cattle, 0.74 million buffaloes, 2.2 million pigs and 23.9 million poultry (FAOSTAT 2010). The growth rates for poultry and pig production are expected to continue at just over 2 per cent annually, only slightly higher than large ruminant production at 1.7 per cent. In sum, the existing population of livestock is already in place to exploit the sector (World Bank 2006, FAOSTAT 2011).

Broadly speaking, livestock sales contribute two-fifths of household income for the poorest 40 per cent of households, while adding one-tenth of household income for the wealthiest 20 per cent of households. In fact, recent surveys have shown that 62 per cent of households keep bovines, 55 per cent keep pigs, and 75 per cent keep traditional poultry (Ifft 2005). With this in mind, there is no doubt that in low-income rural Cambodia livestock is one of the most important sources of easy-access cash income.

However, for all the potential the livestock sector has in lifting people out of poverty, donors and government alike have expressed minimal interest in realizing its true promise. In fact, the state has been a hindrance rather than an enabler for the livestock sector, primarily because no single producer of livestock can yet be a market-maker.

The Ministry of Agriculture, Fisheries and Forestry provides limited technical assistance and poor animal health services delivery to livestock keepers. Government officials have made the transport and official export of cattle and other animal species complicated and onerous to the point that the few livestock exporters have shut their doors (Ear 2005, 2010). Additionally, local critics and detractors cite warnings from international watchdogs saying that raising livestock not only pollutes the environment by adding unwanted greenhouse gases but also consumes large amounts of resources per unit of output (Worldwatch 2005: 14, NGO Forum 2011). On balance, the interaction between external factors – strict and demanding international food markets and the competitive power of neighbouring countries – and the above-mentioned domestic factors has conspired to stunt the growth of the livestock sector (Ear 2005). Only by addressing these inhibiting factors can the sector evolve.

More promising than livestock is the rice sector, but only if rice milling (processing) can emerge and replace large unofficial exports of husked rice to neighbouring countries (primarily Vietnam). The true potential of rice lies in the fact that there are nine million rice farmers in Cambodia, the vast majority of whom live in rural areas. From 2008 to 2010, price developments in rice signal an upward trend: the profitability of the sector has risen. This economic incentive has turned it into an attractive agricultural activity, but the conditions for rice production are somewhat different from livestock: there are hundreds of domestic rice mills that compete, unfairly, with a major state-owned enterprise that has previously aimed to extract rents. Rice milling is a relatively uncomplicated processing task, but to achieve export-quality standards milled rice requires milling equipment that is at the moment not in prevalent use in Cambodia. So, despite sky-high prices in 2008, Cambodia has not yet been able to convert its rice surplus into the white gold it wishes owing to limited international involvement and a small number of regional buyers (particularly Vietnam) willing to pay reasonable prices for husked rice. In addition, there are competing rice millers' associations, and inter-ministerial rivalries between the Ministry of Commerce and the Ministry of Agriculture.

Unlike livestock, variable rice production at household level is flexible and quick enough to respond to price incentives. Also, rural farmers perceive rice farming as less risky and the government sees it as a way to ensure food security. Rice millers, driven in large part by regional and international market prices, turn excess output into immediate profits by buying low and selling high abroad (Thailand and Vietnam) as well as in-country. Because Cambodian rice-milling technology is largely antiquated, those few rice millers who make large capital investments and technological upgrades can offer the highest-quality rice that commands the highest foreign prices. In detecting the existing areas for improvement, private- and public-sector actors in the country can effectively address a number of activities to add value to rice, such as Ziploc packaging.

As expected in countries where governance is weak, in considering the large revenues to be made through unofficial exports, most of the rice business is concentrated jealously in the hands of few to the detriment of many. In sum, the development of agriculture and agro-processing is key for Cambodia's survival in the global economy (World Bank 2004). The choice between exploiting animals or crops will depend much on how internal and external factors play out in relation to national regulations and international incentives against a background of bitter power politics and special domestic interests. If Cambodia seeks to add value to agriculture, it must get its house in order. In the next section we aim to elucidate the challenges encountered with agricultural interventions.

A Bumpy Journey: Nuisance, Potholes and Roadblocks

Interventions, whether national, regional or international, are not immune to problems. A good example is the interminable fight against hunger: it remains an intractable issue no matter how much attention and money is devoted to it. In Cambodia, for example, building the livestock and rice sectors has been fraught with challenges. Some are unique while others are cross-cutting, hence further examination must bear these features in mind. In terms of livestock, this sector suffers from illegal taxation, that is, demands for heavy and frequent informal payments for transport of livestock within the country. This is one way state employees, such as the police, earning abysmally low salaries, complement their incomes.[12] Also, after joining the World Trade Organization in 2004, the production of animals for food and income became an afterthought to senior government officials and line ministers in light of an exploding garment sector.

Furthermore, internally, there are shortages of key production inputs, inaccessibility and poor development of consumption markets, widespread human insecurity, recurrent disease incursions from neighbouring countries, limited technical capacity and services delivery at ministry level, largely deficient skills base at village level, and disinclination towards community actions due to low social cohesion as a result of a Khmer Rouge legacy of distrust (Ear 2005). Externally, there are price shocks, restrictions and trade blockage.

In terms of rice, the sector has several constraints such as limited access to favourable credit, high transportation and storage costs, rain-fed rice with low and variable yields, eroding soils and poor soil nutrition, deficient water management, limited reach of extension services, poor research capacity, inability to export to developed countries owing to high entry barriers, convoluted and expensive logistics, high energy costs and unreliability of electricity delivery. In addition, rents must invariably be shared with the Economic Police during the transport of rice within Cambodia, and if rice must leave the country then customs officials must also be paid at border control points.

Moreover, rice farmers compete with state-owned and state-favoured enterprises in trying to sell outputs at competitive prices. In view of this, one would think that no bright spot exists, but rice may have an opening in exports to the European Union through its zero-tariff initiative under a preferential programme for least developed countries known as 'Everything But Arms'; however, this will require compliance with high sanitary and phytosanitary standards[13] before export permits are granted. Lastly, the overall demand for Cambodian rice is unlikely to be significant given Thailand and Vietnam's dominance in the rice-for-export sector (Ear 2009).

In terms of both livestock and rice, no credible or reputable private-sector organization for collective action exists in either sector, resulting in little bargaining and contestation power in the face of corruption from state agents. Also, because foreign involvement is minimal – when compared to other sectors such as garments – the opportunities for strategic alliances and tactical partnerships to circumvent constraints are closed down.

Animals and crops, both being items of trade, require fossil-fuel-based electricity, diesel or petrol, and transportation, all of which are subject to unofficial payments throughout the state structure. Live animal transport in particular is

---

12   This information was gathered from communication with locals and personal observations in the field.

13   These are standards for animal, plant (including bulbs and tubers, seeds for propagation, fruits and vegetables, cut flowers and branches, cereals and grains) and forest products that deal with agreed levels of chemicals such as bactericides, fungicides, pesticides, antibiotics and growth promoters, as well as presence of diseases or insects. Importing countries only require phytosanitary certificates for some items.

vulnerable, owing to the fact that it is perishable when detained for hours on end under the hot Cambodian sun. Lastly, political stability is a necessary but not sufficient condition for the success of both sectors. The mere fact that local governance is good enough to incentivize garments does not necessarily mean that it is also good enough for livestock and rice, given that the drivers for success are different and require a specific set of conditions to excel as other demand-driven sectors have in the past (Ear 2005, 2009). Soysa and colleagues (1999) see 'good governance as crucial in building healthy conditions for agriculture, and thus in breaking the vicious cycle of poverty, scarcity, and violence'.[14]

In sum, the structural failures of democratic governance in Cambodia stymie the values-based, goals-directed decisions on collective prosperity that need to be sustained over long periods to materialize. It would be wrong, however, to apportion all the blame for lagging agricultural sectors on failing governance. In Cambodia's case there are also elements of myopic and misguided leadership. This is not uncommon in low-income countries hastily inserting themselves into the world economy to fuel rising incomes. The truth is that they are just too busy with the monumental number of day-to-day assignments, obligations and duties associated with economic transitioning processes.

A possible reasoning for this myopia in policymaking could be this: because the benefits of truly sustainable agriculture lie in the future, while interventions must be undertaken now, evidence-based analysis of Cambodia's agricultural potential has thus far proved insufficient to motivate private- and public-sector actions and investments, especially if the discount rate is high. Bluntly put, there are easier and faster routes to move the country forward economically (for example, natural resource concessions, garments for export markets, oil exploration, aid dependence), and agriculture is seldom one of them. This could partly explain why adding value has not been Cambodia's strength.

Now, as in the past, exporting garments and developing tourism have been the most convenient ways to attract and build foreign currencies. But garment factories with their foreign owners arrived almost like ready-to-wear clothes in an empty store, with little incentive for the state to be anything more than a glorified gatekeeper, as Collier has argued (2008), collecting tolls between their national economy and the international system, whether these are official or unofficial (bribe tax), for the trade advantages of a 'Made in Cambodia' least developed country label in foreign markets.

To be fair, other countries have also followed this path (such as Honduras and Tanzania). Of course, this scenario is short-sighted. While evidence-based analysis has not found fertile ground in agriculture, it has struck gold in the politics of poverty: it publicizes catastrophes, disasters, diseases,[15] hunger, illness,

---

14   These views are in agreement with the writings of Ear 2005, 2009 and 2010 in Cambodia.

15   For example, one in-country study notes that 'a rushed, emergency-oriented response by donors to avian influenza may indeed have undermined already weak

poverty and suffering to attract international aid that enriches state gatekeepers. Comparatively, it is a quick fix to inject resources without doing any long-term capacity building. Economic development experts acknowledge that aid in the hands of inept leaders is an intoxicating cure that slows the impetus for urgent internal reforms (Sachs 2006, Easterly 2007).

All in all, significant investments in agricultural education, commodity-transforming factories, affordable national financing, by-product processing facilities, domestic consumption markets, large-scale storage and land and sea transports have been too few and far between to unleash the latent potential of the agriculture sector owing to, among other things, weak state capacity and poor governance. In the next section we reflect on the vicissitudes of Cambodia's experience and the lessons to be learned from them.

Reflections

During and after conflict, the incentives, environment, infrastructure, economy and skills necessary to maintain agricultural production are fragile. In many ways, post-conflict states regress to medieval agrarian societies. The agricultural production capacities, central to the survival and empowerment of citizenries, are much reduced and, in most cases, require rebuilding from the bottom up (Sorpong 1997).

This being the case, post-conflict societies, like Cambodia, can only succeed if there is social reorganization of the three central elements of economic life: production, distribution and exchange. This sequence is only as strong as its weakest element. After conflict, the reconstruction and maintenance of state capacity that is able to exert institutional power through Cambodian society enables the state to formulate and implement policies that advance the security of its citizens through the maintenance of law and order, as well as policies that promote sustainable economic growth and development (Gill 2003: 22–7).

International observers note that for economies heavily reliant on the agricultural sector, one prerequisite to successfully leap into the world economy is the improvement of their agricultural infrastructure, such as irrigation, rural electrification, telecommunications and most importantly, roads. Primary and secondary roads are needed to connect goods to markets (Burgos 2010).

From a strategic viewpoint, agents engaged in conflict tend to destroy infrastructure. While infrastructure, be it agricultural, primary or secondary, is utterly important, its presence will not move a country in the direction of healing its economic, social and political wounds. This is just the beginning of a long journey.

As this case study illustrates, chosen paths in the agricultural realm are fraught with challenging difficulties that can be overcome by an alignment of favourable circumstances to achieve set goals. Very rarely do alignments of this nature occur,

---

governance capacity in Cambodia, fuelling patronage networks and encouraging rent-seeking behaviours' (Ear and Burgos Cáceres 2009: 638).

but state and non-state actors can find common grounds where convergence of interests can shine light on economic sectors and subsectors that can benefit from internal as well as external boosters.[16] For instance, as Ear (2009) notes, external pressures from aid agencies, regional cooperation and economic bodies, and international financial institutions can be instrumental in creating the opportunity for growth in a specific sector.

Furthermore, the multiple roles of international drivers, foreign investors, domestic collective-action mechanisms, and high potential for sustainable revenues are critical to achieve sector growth. In addition to tactical leverage of selected sectors, it is worth suggesting to state officials and private-sector players to stay attuned to global trends and regional developments given that plenty of profitable opportunities are constantly arising in a globalized economy where space and time have been compressed with the assistance of advanced information technology and well-connected international air transport. For example, within Asia, both China and India are now considered emerging market economies where their populaces are enjoying more jobs, better incomes and higher levels of urbanization, all of which can be translated into swelling consumption of goods and services arriving from the most affordable sources. Similarly, within South-East Asia, both Thailand and Vietnam have long been driving regional growth and technological development with the aid of a more educated and vibrant citizenry.

With this in mind, Cambodia is poised to take advantage of growing regional and international demand for livestock, especially when the logistics of trading at shorter distances make the activity ever more attractive. This is easier said than done. Widespread and long-lasting change must come from within, not only from society as active participants in democratic life but also from centrally and bureaucratically organized administrations. Cambodians must be responsible for their own futures.

To this effect, Gill (2003: 27–8) notes that there are two principal means of achieving effective state power over economic affairs: first, by direct involvement in economic activity (for instance, state-owned enterprises), and second, by laws (rules, regulations, standards). In Cambodia, both means have been hijacked by state officials and ruling elites while at the same time intentionally crowding out competitors from high-rent economic sectors. It is at this juncture that constituencies should voice their distress to the international community so that external pressure can be placed on state body restructuring and internal procedural reforms. It is unlikely, however, that these measures will precipitate change. If political unwillingness remains, and weak state capacity persists, donors may consider bypassing the government altogether to work directly with the poor on their own or through non-governmental organizations that have been effective in promoting cost-effective agricultural rural progress (Ear 2005).

---

16   For example, a strategic study of Sino-Cambodian linkages notes that 'China's heavy investment in Cambodia stems not only from Beijing's strategic interests in the region but also arises because private sector initiatives seek fertile ground where there is an environment to prosper and grow' (Burgos and Ear 2010: 638).

In terms of advocacy, opinion leaders and outspoken issues champions need to be reminded that sustainable agriculture is central to effective post-conflict recovery because it provides human security as well as societal stability and rule of law. The importance of agriculture to transitioning economies must be stressed to current and potential international leaders and policy entrepreneurs given that the precipitous descent of some of the world's poorest countries into food insecurity, instability and poverty raises the risk of potentially detrimental spill-over effects, ranging from a rise in illegal migration and organized crime to the encroachment of radicalized religious extremists who might exaggerate their grievances abroad to attract attention (Burgos 2010, Burgos and Otte 2011).

Finally, in Cambodia, while the challenge of adding value to agriculture after conflict is in part a result of poor governance, lack of commitment, corruption, overt elitism and opportunism, there are numerous strengths and opportunities present to leverage agriculture and agricultural processing so that this sector can be positioned right at the top of both the economic and political agenda given its immediate and widespread potential to lift large swaths of the population out of poverty.

The lesson to be learned from the Cambodian post-conflict experience regarding value addition to agricultural products is that strategic advantages and political will play a pivotal role in determining which items get pushed domestically or internationally in terms of attention, funding, technical assistance, and support in the marketplace. It is also important to recognize that countries wishing to attain quick wins are going to opt for sectors generating high profits and informal revenues (bribe taxes). Agriculture requires a long-term outlook and is weather-dependent. In other words, the risk of failure or underperformance in long-term agricultural investments is very high, so politicians with short-term outlooks, who fear losing power, will favour quick wins in non-agricultural investments.

It can therefore be argued that conflicts are double-edged swords: they bring death, destruction, misery, and are, as Paul Collier calls wars, development in reverse, but at the same time they raze the playing field for all possible agricultural items to have a shot at becoming leading export products.

References

Burgos Cáceres, S. 2010. Africa's woes, world's opportunities. *International Journal for Rural Development*, 44(6), 33.

Burgos Cáceres, S. and Ear, S. 2010. China's strategic interests in Cambodia: influence and resources. *Asian Survey*, 50(3), 615–39.

Burgos Cáceres, S., Hinrichs, J., Otte, M.J., Pfeiffer, D., Roland-Holst, D., Schwabenbauer, K. and Thieme, O. 2008. *Poultry, HPAI and Livelihoods in Cambodia – A Review*. Working Paper No. 3. Rome: Food and Agriculture Organization of the United Nations.

Burgos Cáceres, S. and Otte, M.J. 2011. Animal health and international affairs: trade, food, security, and global health. *Yale Journal of International Affairs*, 6(1), 108–9.

Burgos Cáceres, S. and Otte, M.J. 2009. Blame apportioning and, B. and Otte, M.J. 2009. Blame apportioning and the emergence of zoonoses over the last 25 years. *Transboundary and Emerging Diseases*, 56(9–10), 375–9.

Center for Stabilization and Reconstruction Studies (CSRS). 2010. *Agriculture: Promoting Livelihoods in Conflict-Affected Environments*. Workshop Report, Center for Stabilization and Reconstruction Studies. Monterey, CA: US Naval Postgraduate School.

Chandler, D. 2007. *The History of Cambodia*. 4th Edition. New York: Westview Press.

Colleta, N.J. and Cullen, M.L. 2000. *Violent Conflict and the Transformation of Social Capital: Lessons from Cambodia, Rwanda, Guatemala, and Somalia*. Washington, DC: World Bank.

Collier, P. 2008. *The Bottom Billion: Why the Poorest Countries are Failing and What Can Be Done About It*. New York: Oxford University Press.

Collier, P., Elliot, V.L., Hegre, H., Hoeffler, A., Reynal-Querol, M. and Sambanis, N. 2003. *Breaking the Conflict Trap: Civil War and Development Policy*. Washington, DC and New York: World Bank and Oxford University Press.

Collier, P. and Pattillo, C. 2000. *Investment and Risk in Africa*. New York: St. Martin's.

Ear, S. 2005. *The Political Economy of Pro-Poor Livestock Policy in Cambodia*. PPLPI Working Paper 26. Rome: Food and Agriculture Organization of the United Nations.

Ear, S. 2009. *Sowing and Sewing Growth: The Political Economy of Rice and Garments in Cambodia*. SCID Working Paper 384. Stanford, CA: Stanford Center for International Development.

Ear, S. 2010. Cambodia's patient zero: global and national responses to highly pathogenic avian influenza, in *Avian Influenza: Science, Policy and Politics*, edited by I. Scoones. London: Earthscan.

Ear, S. and Burgos Cáceres, S. 2009. Livelihoods and highly pathogenic avian influenza in Cambodia. *World's Poultry Science Journal*, 65(4), 633–40.

Easterly, W. 2007. *The White Man's Burden: Why the West's Efforts to Aid the Rest Have Done So Much Ill and So Little Good*. New York: Penguin Press.

FAOSTAT. 2010. *Production: Live Animals*. [Online]. Available at: http://faostat.fao.org/ [accessed: 30 November 2010].

FAOSTAT. 2011. *Production: Livestock Primary*. [Online]. Available at: http://faostat.fao.org/ [accessed: 13 May 2011].

Food and Agriculture Organization of the United Nations (FAO). 2007. *Gridded Livestock of the World 2007*. Rome: FAO Press.

Food and Agriculture Organization of the United Nations (FAO). 2010. *Livestock Sector Policies and Programmes in Developing Countries: A Menu for Practitioners*. Rome: FAO Press.

Gill, G. 2003. *The Nature and Development of the Modern State*. Basingstoke: Palgrave.

Gore, A. 2009. *Our Choice: A Plan to Solve the Climate Crisis*. Emmaus: Rodale.

Ifft, J. 2005. *Survey of the East Asia Livestock Sector*. Working Paper, Rural Development and Natural Resources Sector Unit of the East Asia and Pacific Region. Washington, DC: World Bank.

International Livestock Research Institute (ILRI). 1995. *Livestock Policy Analysis*. Nairobi: ILRI Press.

International Monetary Fund (IMF). 2010. *Cambodia and the IMF*. [Online]. Available at: http://www.imf.org/external/country/KHM/index.htm [accessed: 5 December 2010].

McKibben, B. 2007. *Deep Economy: Economics as if the World Mattered*. Oxford: Oneworld Publications.

Marks, P. 2000. China's Cambodia strategy. *Parameters*, 30(3), 92–108.

Modell, J. and Haggerty, T. 1991. The social impact of war. *Annual Review of Sociology*, 17, 205–24.

NGO Forum on Cambodia. 2011. *Environment Program*. [Online]. Available at: http://www.ngoforum.org.kh/Environment/Env-Pro.htm [accessed: 13 May 2011].

Sachs, J. 2006. *The End of Poverty: Economic Possibilities for Our Time*. New York: Penguin Press.

Sorpong, P. 1997. *Conflict Neutralization in the Cambodia War: From Battlefield to Ballot-box*. Kuala Lumpur and Oxford: Oxford University Press.

Soysa, I., Gleditsch, N.P., Gibson, M. and Sollenberg, M. 1999. To cultivate peace: agriculture in a world of conflict. *Environmental Change and Security Project Reports*, 5(summer), 15–25.

World Bank. 2004. *Cambodia at the Crossroads: Strengthening Accountability to Reduce Poverty*. East Asia and the Pacific Region, Report No. 30636-KH. Washington: World Bank.

World Bank. 2006. *Cambodia: Halving Poverty by 2015? Poverty Assessment 2006*. Washington: World Bank.

World Bank. 2010. *Cambodia: Data and Statistics*. [Online]. Available at: http://web.worldbank.org/WBSITE/EXTERNAL/COUNTRIES/EASTASIAPACIFIC EXT/CAMBODIAEXTN/0,menuPK:293881~pagePK:141132~piPK:141109~t heSitePK:293856,00.html [accessed: 5 December 2010].

Worldwatch Institute. 2005. *Vital Signs 2005*. Washington, DC: Worldwatch Institute.

Chapter 11

Avoiding Dairy Aid Traps: The Cases of Uganda, India and Bangladesh

Bruce A. Scholten and Brian T. Dugdill

Introduction

Dairying plays versatile roles in the nutrition and livelihoods of smallholders and their communities. Animals such as cows, buffalo, camels, goats, horses, reindeer and sheep (henceforward 'cattle') convert plants inedible to people into high-protein foods. Cattle are stores of wealth where banks do not function, providing fertilizer and draught power for ploughs and carts. In mature dairy economies such as the USA and UK, most milk animals exist in intensive confined animal feeding operations requiring high energy use; in our case studies of Uganda, India and Bangladesh, around 80 per cent of dairy sectors are informal, and many smallholders keep one or two cows for family food – but have surplus milk that could be sold (Scholten 2010a, 2010b).

Compared with the USA and UK, where labour is shed by high technology, dairying in developing countries is generally labour-intensive, involving family members in milking and feeding. A landless widow may keep a cow, which grazes roadsides and returns home at nightfall for milking and a ration of grain. Importantly, dairying in developing countries provides one long-term full-time off-farm job for every 20–50 litres of milk collected, processed and marketed (Omore et al. 2004). For such reasons the United Nations Development Programme and Food and Agricultural Organization (UNDP-FAO) promote dairy marketing systems to raise subsidiary incomes of small-scale farmers in peacetime or rehabilitate countries after war, by improving rural-to-urban cold chains (Dugdill et al. 2000).

Short-term food aid may be needed to save lives in war or natural emergencies. But in the medium and long term, it is important to avoid food donations that become *aid traps* destabilizing previously sustainable agricultural sectors in recipient countries. Surplus disposal from rich countries masked as food aid can harm recipient country markets, if arrival overwhelms local harvests. The assumed benefits of food aid may also be negated if aid is tied to imports of unneeded or inappropriate technology. In developing areas, petroleum-dependent high tech may be appropriate only at mass processing levels, while laborious hand-milking, simple stainless steel milk cans and insulation – rather than costly refrigeration – are best to link farm to consumer (Cheruiyot 2011).

Following regional wars and the violent tenure of Idi Amin, 1971–9, Uganda was dependent on commercial and donated dairy commodities. Thus, this chapter's centrepiece is the UN multi-donor dairy rehabilitation project in Uganda in the 1980s. It is compared with the UN's less effective Bangladesh cooperative dairy development project in the 1970s, after its war of liberation from Pakistan. Further lessons are drawn from history's largest dairy aid programme, India's successful Operation Flood, which tripled milk production during a period when global production rose by half (*Economist* 2011). Before presenting these case studies we will discuss dairy aid in more detail.

Food Aid in Agricultural Instability

Among the factors that can destabilize previously sustainable agricultural sectors are military conflict, colonial mismanagement, unequal trade relationships and natural disasters. Food aid can be a partial solution for emergencies, such as the 2004 Indian Ocean tsunami. But if a scenario is the result of decades of agricultural instability caused by war or poor governance, there may be a case for long-term food aid programmes, to resuscitate food chains and – especially in countries with large farm sectors – to foster livelihoods and maximize incomes of large numbers of families.

But food aid brings its own risks to food chains, which can be disrupted by corruption in supply chains, unviable price controls, inept monetization of aid commodities or – the focus here – what amounts to *dumping*. In war or peace, food networks cannot operate if actors or commodities are not in place. Misappropriated or untimely commodity deliveries effectively equate to dumping. This practice, of exporting goods to another country at less than normal value, is so destructive to indigenous farmers that the 1994 Uruguay Round Agricultural Agreement (URAA), signed with the establishment of the World Trade Organization, is also known as the 'Uruguay Round Anti-dumping Agreement' (Bekker 2005: 9).

Dumping can stem from a mix of good intentions and commercial or national needs for surplus disposal. Hoekman and Kostecki (1995: 176) write of 'cyclical' and 'strategic' dumping. European Economic Community (EEC) aid was partly triggered by the high costs of storing dairy commodities, and political inability to cut farm subsidies.

Hans Singer is a figurehead for practitioners such as Eddie Clay, James Ingram, John Shaw and John W. Mellor of the World Food Programme (WFP), advocating food aid to ease distress from war, or structural adjustment in peacetime. Singer (1991) urges recipients to minimize disincentives to farmers, noting 'the real limit is not aid capacity, but absorptive capacity'. Food aid has saved many lives in post-conflict situations, but even in emergencies such as famines, doubters rightly ask: Will aid be a disincentive to farmers, skew markets and create dependence? (Cunningham 2009: 11, Doornbos and Gertsch 1994). The disincentives of food aid are in a precarious balance with its potential. A food system functions well in equilibrium, but any hole in it invites failure. When aid is counterproductive, the 'smoking gun' often turns out to be commodity dumping.

Case of India

Although India's White Revolution, including the Operation (milk) Flood dairy programme, of 1970–96 transpired mostly in peacetime, it began shortly after border incidents with China, and continued during chronic tension with Pakistan. It was the biggest-ever programme involving dairy aid from the European Economic Community (EEC) facilitated by UN-FAO, the World Food Programme and the World Bank (Candler and Kumar 1998, Scholten 1998, 2010a). After 1946, when a farmers' strike dislodged the British-licensed monopoly that supplied Mumbai with milk, the subsequent farmers' dairy cooperative near Anand, Gujarat, was famed for its brand name Amul, all collectively dubbed the Anand Pattern. Milk prices to farmers were higher in Operation Flood than under the Raj, and this increased the share of dairying in gross domestic product (GDP) due to widespread participation in dairying by marginal and landless farmers. The crucial point is that the National Dairy Development Board (NDDB, a parastatal government institution sheltered by four presidents from political meddling at its headquarters far from Delhi in Anand) improved India's food security – and food sovereignty – by managing food aid without disincentivizing farmers. Kenda Cunningham (2009) writes that the NDDB sought to replicate the Anand Pattern but 'The EEC had been donating skimmed milk powder and butter to India since the late 1960s and Anand allegedly viewed this as a threat to its market.'

The threat to India's dairy sovereignty was real. Shanti George (1985: 249) warned that EEC dairy gifts could be 'Trojan horses'. Peter J. Atkins (1988) cites EEC estimates of dairy aid to Operation Flood of 441,000 metric tonnes of skimmed milk powder and 151,000 metric tonnes of butter oil. Such quantities could have made dairy cooperatives unsustainable in many areas of the country, bringing permanent import dependence.

Hans Singer (1991: 140) maintains that where food aid is warranted it is 'subject to assurance that food aid helps to provide incentives for increased local production and physical and human investment in recipient countries'. So, the greatest test of Operation Flood was whether India could wean itself from European dairy aid and boost its own production in the Anand Pattern. In India's federal political structure, individual states are responsible for agriculture; thus state politicians sometimes negotiated aid deals with foreign donors that disincentivized local farmers. This helps explains why national milk production actually fell below 20 million metric tonnes in the 1960s. Banerjee (1994) presages Cunningham above, noting that Operation Flood's architect Verghese Kurien and the NDDB heard reports of mounting dairy surpluses in Europe as warnings for India's struggling milk farmers, but they turned crisis into opportunity, by winning government support in 1970 to establish the Indian Dairy Corporation (IDC) to manage monetization of all EEC commodities. This brought holistic order to foreign supply. Monetization enabled Anand to treat aid as additional to domestic milk, and its proceeds funded a national milk grid infrastructure, balancing foreign and domestic supply among Mother Dairies in Delhi, Kolkata, Madras and Mumbai. Led by Anand, this became known as India's White Revolution.

Scholten (2010a: 23, 182, 210, 220–27, 244) claims India sidestepped any potential European dairy aid trap in 1984, when the Jha Commission Report (Jha et al. 1984) strengthened Anand's control of aid – and guaranteed the success of Operation Flood – by mandating that processors pay more for EEC commodities than domestic milk. In post-1945 Germany food shortages vanished when Allied authorities revoked price controls (Mee 1984). Similarly, when India's Mother Dairies offered better prices to Indian farmers, milk flooded from villages, quintupling production by 2000, and nearly doubling per capita dairy consumption amid rapid population growth. From a cooperative of a few hundred Gujarati farmers in 1946, the Anand Pattern attracted over 10 million members in the next six decades. India forsook EEC aid, passed the USA as top world milk producer, and exports to over 40 countries (*Financial Times* 2008). The World Bank (2008b, 2009) now promotes the Anand Pattern in Africa, although so far with limited success.

Case of Bangladesh

After its war of independence from Pakistan, the dairy programme in Bangladesh was designed to coordinate domestic dairying with targeted imports of foreign dairy commodities (Shah and Dugdill 1979). The government faced an acute milk shortage and called in the UN for its first major agricultural development effort (Haque 2008). Milk was selected as one key commodity to drive rehabilitation because it provides nutrition for the poorest farming families and generates regular cash earnings. Technical and financial aid came from UNDP, FAO and the Danish International Development Agency via the government, with the goal of establishing sustainable dairy development by setting up the Bangladesh Milk Producers' Cooperative Union Limited – known by its Milk Vita brand – adapted from India's Anand Pattern. Dugdill et al. (2000) recall:

> From a modest start … providing 4,300 very poor, often landless, households in remote rural areas with a complete package of milk production enhancing technologies, village level organisational skills and a milk collection-processing-marketing system, the two-tier Milk Vita Co-operative has grown into a successful commercial dairy enterprise. Today, milk is collected from 40,000 farmer-members organised into 390 primary village co-operatives, then processed and distributed to all the major cities in the country. In 1998, producers delivered 30 million litres of milk surplus to their household requirements and received a total of Taka 467.4 million (US$ 9.3 million). … This has helped to

lift average earnings above the poverty line and to sustainably improve rural livelihoods.

Dugdill et al. (2000) note that urban deliveries and milk freshness were improved via locally fabricated three-wheeled insulated milkshaws and that 'one off-farm job was created for each 35 litres of milk collected, processed and marketed. More than half the jobs are in rural areas.' Sustaining farm and off-farm incomes above the poverty line helps citizens meet the challenge of monsoon flooding. Much effort was expended by the government of Bangladesh and entities such as the FAO to free routes to market to enable rural producers to ship raw milk to processors and the great conurbation of Dhaka. Rural village milk-producing cooperatives supplying high-quality milk were linked directly by Milk Vita to urban milkshaw cooperatives distributing safe and affordable milk. Today, three decades on, Milk Vita has many imitators and the dairy value chain remains a huge income and job generator.

But Milk Vita faced the classic danger of food aid: dumping. Television images of hungry children in Bangladesh generated demands in the rich world that something must be done. Global viewers became actors in conflict resolution. Butter oil and milk powder from outside the UN dairy programme was shipped from the EEC, but much of this 'aid' arrived too late to ease hunger. Bangladesh's milk market was approaching effective equilibrium when a large shipload of EEC dairy aid docked.

Local dairy project authorities objected to the commodities, saying they were unfit for human consumption – having sailed through a typhoon in the Bay of Bengal during which sea water leaked into the hold, which also contained urea fertilizer stowed on top of the milk powder. But they also feared the commodities posed a severe threat to tenuous price structures. Their alarm was not spurious. Soon well-connected dealers obtained control of this 'aid'. In weeks, tonnes of milk powder and butter oil were sold in Bangladeshi black markets – at prices below the cost of milk production in Europe – wrecking much of the progress that had been made, and rendering local dairy chains unsustainable. Blame fell on Europe.

The FAO Terminal Report for the programme (Dugdill 1986) relates how these 'donated' dairy commodities, and other so-called 'commercial' imports of full-cream milk powder originating mainly from Europe, were in fact commodities subsidized at a rate of 50–60 per cent of domestic prices in exporting countries (Dugdill 1986). The FAO chronicles how milk reconstituted from imported dairy commodities retailed at a liquid-equivalent price of just 20 US cents per litre, compared with locally produced pasteurized milk at 38 US cents. Between 1982 and 1986, milk powder imports increased fourfold to 66,385 tonnes annually, equivalent to about 56 million litres of liquid milk. This unfair competition resulted in Milk Vita's annual share of the market collapsing from 10 million to 3 million litres. This was not because the Milk Vita value chain was inefficient, or unsustainable, as just before the dumping farmers received 60 per cent of the retail pasteurized milk price – well above the unsubsidized share received by farmers in the West.

Dumping decreased after 1984 when Europe imposed milk quotas to reduce surpluses (Scholten 1990). A quarter-century later Milk Vita is such a 'flourishing concern' that 'imitators ... set up similar dairy enterprises to process and market 50 million litres of milk annually', and the Grameen Bank added dairying to its development model (Dugdill et al. 2000: 11). In his book *Creating a World without Poverty* (2008), Nobel Laureate Muhammad Yunus explains how the Grameen Fisheries and Livestock Foundation helps women participate in social businesses involving crops-dairy-fish farming.

Nevertheless, cooperative dairying in Bangladesh has not really equalled the success of India's Anand Pattern. FAO data show the rate of exponential growth of butter and ghee per capita consumption between 1961 and 1989 in Bangladesh was negative (−2.4 per cent) while India's was positive (+0.7 per cent), according to Scholten (2010a). Reasons for this are complex, related to conflict, poverty and a huge surge in population. But the lesson is that farmer incomes probably would have improved years earlier if Bangladeshi authorities had been able to exert better control of dairy aid from abroad.

Case of Uganda

Uganda's dairy industry rehabilitation programme in the 1980s was more successful than Bangladesh at post-conflict resolution by satisfying hunger and raising farm incomes. Balikowa (2011) notes that Uganda's dairy sector has developed rapidly in recent years, but remains dominated by 'small-scale farmers owning more than 90 per cent of the national cattle population'. The dairy history of this country of 35 million people has contemporary relevance because Uganda is a member of the East Africa Dairy Development (EADD) project with Kenya and Rwanda, with expansion to Tanzania and Ethiopia scheduled for 2012.

Established in 2008, the EADD pilot phase is funded by the Bill and Melinda Gates Foundation through the not-for-profit US organizations Heifer International and TechnoServe, as well as the international organizations ILRI (2011, International Livestock Research Institute) and ICRAF (World Agroforestry Centre), and private-sector bodies such as ABS-TCM (African Breeding Services – Total Cattle Management). Multinational companies such as Nestlé and Tetra Pak now partner with the project. EADD is being piloted in 68 regional dairy hubs: 21 in Kenya, 16 in Rwanda and 31 in Uganda. The overall goal of the pilot phase, which runs from 2008 to 2012, is to transform the lives of 179,000 families, or about 1 million people, by doubling household dairy income in ten years through integrated interventions in dairy production, market access and knowledge application. In the main phase between 2012 and 2018, 350,000 more farm families will be added, totalling 529,000 or about 3.2 million people altogether (Cheruiyot 2011, Gaitano 2011).

After decades of conflict and even genocide, EADD is an attempt to reduce poverty in the region by linking commercial, technical and input service provision directly to farmer-owned milk bulking businesses. Prospects improved in 2010 when Burundi, Kenya, Rwanda, Tanzania and Uganda formed the East Africa Community (EAC), planning a free trade area by 2015. Nor is EADD the only dairy force in East Africa. The World Bank (2008b, Scholten and Basu 2009) intends to replicate India's Anand Pattern in Africa, and established a base in Kenya to encourage market solutions with USAID (US Agency for International Development) and Land O'Lakes Cooperative.

A politically forced ethnic diaspora, and domestic and external territorial conflicts, destabilized dairying in the 1970s and 1980s. Uganda became independent from Britain in 1962, with Milton Obote as prime minister. Idi Amin usurped Obote in a 1971 coup, and the next year Amin exiled about 40,000 Asians (*Economist* 2003). Because many worked in the civil service, formal manufacturing and retail sectors – including food and dairy products – their exit also disrupted the lives of Ugandans left behind.

During Idi Amin's rule, perhaps 300,000 Ugandans were killed, many in territorial skirmishes with Tanzania. A Ugandan–Kenyan dispute weakened Amin's military and economic resources, but it was his 1978 invasion of Tanzania that precipitated his downfall; in 1979 Tanzania invaded Uganda, and exiled Amin. Eventually, the new president Yoweri Museveni marked the beginning of Ugandan recovery (*Economist* 2003).

Recovery

Uganda's imports of dairy commodities surged toward the end of the Idi Amin era, due in part to rebel no-go areas in the west, the traditional milk-producing region. This marked a socio-economic crisis, since the loss of income by rural suppliers to the domestic dairy supply depressed the economy – while imports, mainly from more stable Kenya, exacerbated the national trade deficit. Dairy revival was vital to rehabilitation and stabilization efforts. President Museveni, himself an avid breeder of Uganda Ankole cattle, bartered half his country's first coffee export crop in 1987 for 4,800 in-calf Friesian heifers from Europe. Fewer than 2,000 animals were eventually imported, but they laid the foundation for modern dairying (Habyarimana and Dugdill 1991).

Like India, Uganda is an example of how food aid from Europe and WFP has been used effectively to rehabilitate and expand the dairy industry. Government, local stakeholders and development partners, led by the Ministry of Agriculture, Animal Industry and Fisheries (MAAIF) and the UN-FAO, used pasteurized milk, recombined from donated butter oil and skimmed milk powder, to fill the gap between supply and demand in urban areas. Donated commodities were sold to Dairy Corporation – the state-owned dairy company charged with implementing revival (finally privatized in 2006). A rigorously imposed formula was used that ensured the raw material price was always 10 per cent higher than the local farm gate price to foster local milk production. The proceeds from sales were transferred to the Dairy Development Fund, managed by the Dairy Development Committee,

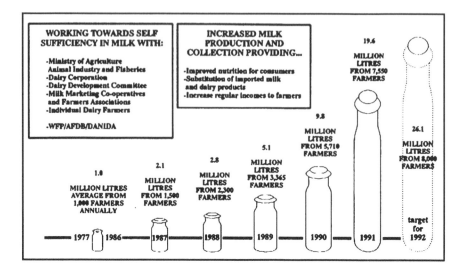

Figure 11.1   Growth in milk production in Uganda (1977–92)

Source: Borland and Moyo 1996. GOVT/UNDP/FAO/UGA/84/023/Dairy Industry Development Project Uganda 1977–1992. Used with permission.

a group of committed public- and private-sector dairy actors. The committee acted both as an incubator for dairy development and a launch pad for investments along the entire cow-to-consumer dairy value chain. In all, over the period 1987 to 1992, some US$25 million was monetized and invested (Dugdill 1992).

Some of the achievements of this period of intense activity include: (i) the main Dairy Corporation plant in Kampala refurbished and its daily throughput increased to 150,000 litres of milk; (ii) 64 milk-collecting centres established and/ or revived; (iii) milk collection increased from 235,000 litres in 1986 to 26 million in 1992 (see Figure 11.1); (iv) donated butter oil and skimmed milk powder monetized through the Dairy Development Committee; and (iv) the Entebbe Dairy Training School and processing plant reopened (Habyarimana and Dugdill 1991). The Dairy Development Committee with its residual funds was the forerunner of the Dairy Development Authority (DDA), set up by an Act of Parliament in January 2002 as developer and regulator of the dairy industry.

But practitioners recall Uganda's dairy renaissance was not without peril. The post-war land was fraught with migrating groups of rebels or ex-rebels, searching for sustenance. Cattle rustling by the Karamojong flourished on a grand scale – approaching 750,000 head in 1988 according to the World Bank (MAAIF 1993). Hijacking of tankers and trucks was common. Initially no banks operated in rural areas. Huge amounts of cash payments for milk were hidden in milk cans, spare tyres and other supposedly safe places in attempts to thwart bandits.

Other incidents had a bizarre side. In 1988 the Dairy Corporation's main Kampala dairy was closed for two weeks because pasteurized milk tested at Makerere University was reported to contain high levels of radioactivity. The milk, recombined from European butter oil and milk powder, was feared contaminated by fallout from the 1986 Chernobyl disaster. Local water at the dairy later tested radioactive. It transpired that the test results were incorrect – a laboratory technician forgot to deduct background radiation. In due course, milk samples sent for analysis at the International Atomic Energy Agency in Vienna proved negative and operations resumed.

The Future

Much of the north and east of Uganda was not included in early revival efforts due to insurgency, first from Alice Lakwena's Holy Spirit Movement, and later from Joseph Kony's Lord's Resistance Army. Activities were initially severely restricted in the infamous Luwero Triangle area in Central Region. The first Dairy Master Plan developed in the early 1990s included northern and eastern milksheds, but little was achieved due to insecurity (MAAIF 1993). Now that the country is more secure, the current National Dairy Strategy and Investment Programme, again being developed with UN-FAO support (FAO 2010), plans to invest US$150 million in dairying over the period 2011–15, with concerted interventions in the northern and eastern Karamoja milksheds (MAAIF/DDA/FAO 2011). About 82 per cent of Ugandans remain in agriculture, and the MAAIF ranks dairy fourth of ten commodities strategically important to the country's Development Strategy and Investment Plan (2010–15). The first three are maize, coffee and fish.

Notwithstanding the challenging past quarter-century, the dairy sector today plays a prominent role in the livelihoods of approximately 1.7 million farming families, close to 10 million people, or almost one-third of the entire population. With its vast areas of pastureland and livestock wealth, Uganda has a huge regional and international agroecological comparative advantage in dairying. The cost of milk production, especially in the south-west, competes favourably with the most efficient developed exporting countries. The dairy industry contributes over half the output of the livestock subsector (Balikowa 2011). The National Dairy Strategy aims at translating these opportunities into nutrition and prosperity for all Ugandans, especially commercializing smallholder dairy farming and ensuring safe dairy products affordable to all (Dugdill 2011). The World Food Programme's positive experience with dairy commodities in Uganda laid the ground for reorientation of its food aid programming and procurement strategy. One of the very first procurements of local food aid in the early 1990s was maize grown in Uganda for the Operation Lifeline Sudan post-conflict emergency. As high-energy maize is key to dairy inputs, this was a positive forerunner of the current WFP P4P (Purchase for Progress) programme now piloted with the Gates Foundation (MAAIF/DDA/FAO 2011).

156 *Challenging Post-conflict Environments*

Conclusions

An environment of government enablement, including appropriate policy support, is necessary for best practice in stabilization agriculture. Getting the prices right is shorthand for balancing incentives for actors from farmers to consumers to act in equilibrium. We have seen how disequilibrium results: a bumper harvest can lower prices for any commodity, and unrealistically low price controls discourage farmers, processors and retailers. But food aid, a factor almost unique to the last century, can operate in capricious ways – like dumping – with harmful consequences for recipients that were unintended by donors. The cases of India, Bangladesh and Uganda show commonalities in the difficulties of sustaining agriculture in war and peace. In particular, aid must be channelled in ways that do not harm indigenous farmers' livelihoods.

The story of India's White Revolution is a quarter-century lesson on the main points of this piece: national food security in disequilibrium situations is linked to control of food aid flows. The first phase of Operation Flood found modest increases in production, partly due to farm extension and crop wastes from the Green Revolution. Great success was attained after 1984, when the Jha Commission forced processors to pay more for EEC dairy aid than domestic milk. Once India got the prices right, milk flowed from villages and revived the dairy sector, which increased its share of national output 'from about 5.6 percent GDP in 1980–81 ... to nearly 6 percent in 1994–95' even as agriculture declined overall and industry boomed (Scholten 1998: 212).

The case of Bangladesh revealed the smoking gun of uncontrolled commodity aid, effectively dumped in the middle of the young country's post-war dairy revival. The damage from untimely aid stymied the dairy sector and the socio-economic well-being of its farmers. It is tantalizing to speculate on the counterfactual of how much quicker the Milk Vita cooperative model might have developed if the Bangladeshi government had been able to exert more control over commodity imports.

This chapter's central case of post-conflict Uganda, where aid did not prove disruptive, is a useful test of the smoking gun theory. Perhaps the geographic aspect of Uganda's landlocked distance from the Indian Ocean offered protection from sea deliveries of aid that shattered price structures in post-war Bangladesh. The general inference is that dairy project authorities in Uganda, like India, avoided interference from foreign commodities, while using them to benefit their own farmers and consumers.

Dairy cattle produce nature's most complete food, providing a valuable source of affordable human nutrition, especially protein and micronutrients, along with animal traction and regular income for farmers. Dairying can contribute significantly to economic stabilization, but aid traps must be avoided. Today's programmes in India, and in East African countries such as Uganda, are on a serious scale and are a testament to earlier pioneering development efforts. They will improve the well-being of many more people – hundreds of millions of rural farming families and urban consumers alike. Even in Bangladesh where progress was slowed by dumping, scaling up is now under way with new public- and private-sector interventions, including those currently funded by the Grameen Bank and the Gates Foundation.

Dairy aid programmes in Bangladesh following its war of independence, in Uganda after war and misrule by Idi Amin, and in post-colonial India brought stabilization *to* agriculture and *through* agriculture, in that they sought a status quo ante, a more sustainable – or sovereign – food system that existed before. The promotion of dairying through contemporary peacetime projects, in East Africa, Bangladesh and India, reflects a realization by experts that small-scale dairying can be a powerful trigger for sustainable and stable socio-economic development, on and off farms.

Acknowledgements

Thanks to colleagues who helped make this piece possible, and to anonymous reviewers for useful advice. Any errors are ours, and constructive comments are welcome.

References

Atkins, P.J. 1988. Rejoinder: India's dairy development and Operation Flood. *Food Policy*, 13(3), 305–12.

Balikowa, D. 2011. *A Review of Uganda's Dairy Industry for the FAO Technical Cooperation Programme Project: Development of an Updated National Strategy for the Dairy Sector and Dairy Value Chain Development* (TP/UGA/3202D). Rome: United Nations Food and Agricultural Organization (UN FAO).

Banerjee, A. 1994. Dairying systems in India. *World Animal Review*, 79. Available at: http://www.fao.org/docrep/T3080T/t3080T07.htm [accessed: 20 May 2011].

Bekker, D. 2005. The strategic use of anti-dumping in international trade. *South African Journal of Economics*, 74(3), 501–21.

Borland, P. and Moyo, H. 1996. Section II: Country papers on dairy development and status of the dairy sector, in *Regional Exchange Network for Market Oriented Dairy Development*, edited by M. Matanda. Rome: FAO. Available at: http://www.fao.org/DOCREP/004/W3199E/W3199E00.HTM.

Candler, W. and Kumar, N. 1998. *India: The Dairy Revolution: The Impact of Dairy Development in India and the World Bank's Contribution*. Washington, DC: World Bank.

Cheruiyot, A. and East Africa Dairy Development project team. 2011. *Unlocking Value of Smallholder Dairy through Hub Model – Case of Kenya*, 7th African Dairy Conference and Exhibition, Dar es Salaam, 25 May 2011.

Cunningham, K. 2009. Smallholder dairy: Operation Flood in India, in *Millions Fed: Proven Successes in Agricultural Development*, edited by D.J. Spielman and R. Pandya-Lorch. Washington, DC: International Food Policy Research Institute.

Daniel, R.N.T. 2008. *Agricultural Innovation and Technology in Africa: Rwanda Experience: Coffee, Banana and Dairy Commodity Chains*. World Bank Institute.

Doornbos, M. and Gertsch, L. 1994. Sustainability, technology and corporate interest: resource strategies in India's modern dairy sector. *Journal of Development Studies*, 30(4), 916.

Dugdill, B.T. 1986. *Terminal Report: Project Findings and Recommendations: FAO/UNDP Cooperative Dairy Organisation and Extension Programme, Bangladesh* (BGD/79/033). Rome: FAO.

Dugdill, B.T. 1992. *Terminal Report: Project Findings and Recommendations: FAO/UNDP Dairy Industry Development project, Uganda* (UGA/84/023). Rome: FAO.

Dugdill, B.T. 2011. *Guidance Notes for MAAIF/DDA/FAO National Dairy Strategy*. Rome: FAO.

Dugdill, B.T., Bennett, A., Saha, G.C. and Das, S.C. 2000. *Milk Vita in Bangladesh – FAO*. Rome: FAO.

Economist. 2003. Amin's legacy in Uganda: bloody tyrant, now a good sort: memories of Uganda's late despot are surprisingly short. *Economist*, 21 August.

Economist. 2011. The only reliable way to produce more food is to use better technology. *Economist*, 24 February.

Financial Times. 2008. Amul exports to 40 countries. *Financial Times*, 9 June.

Food and Agricultural Organization of the United Nations (FAO). 2010. *FAO Technical Cooperation Programme Project: Uganda Development of an Updated National Strategy for the Dairy Sector and Dairy Value Chain Development* (TCP/UGA/3202D). Rome: FAO.

Gaitano, S. 2011. *The Design of M&E Systems: A Case of East Africa Dairy Development Project*, INTRAC 7th Monitoring and Evaluation Conference, Netherlands, 14–16 June.

George, Shanti. 1985. *Operation Flood: An Appraisal of Current Indian Dairy Policy*. Delhi: Oxford University Press.

Habyarimana, J.N. and Dugdill, B.T. 1991. *Uganda Country Paper, FAO Consultation on Dairy Development*. Rome: FAO.

Haque, S.A.M.A. 2008. Bangladesh: social gains from dairy development, in *Smallholder Dairy Development: Lessons Learned in Asia*, edited by N. Morgan. Rome: FAO.

Hoekman, B. and Kostecki, M. 1995. *The Political Economy of the World Trading System: From GATT to WTO*. Oxford: Oxford University Press.

Jha, L.K., Rau, S.K., Bhatty, I.Z., Dastur, N.N., Bhattacharya, P. and Shirali, A.R. 1984. *Report of the Evaluation Committee on OF-II*. Delhi: Ministry of Agriculture, Government of India.

MAAIF. 1993. *Master Plan for the Dairy Sector*. Kampala and Rome: (Uganda) Ministry of Agriculture, Animal Industry and Fisheries, DANIDA and FAO.

MAAIF/DDA/FAO. 2011. Draft National Dairy Strategy 2011–2015. Kampala and Rome: MAAIF/DDA/FAO.

Mee, C.L. 1984. *The Marshall Plan*. New York: Simon & Schuster.

Njehu, A., Omore, A., Baltenweck, I. and Muriithi, B. 2011. Livestock disease challenges and gaps in delivery of animal health services in East Africa. Nairobi/Addis Ababa: International Livestock Research Institute.

Omore, A., Cheng'ole Mulindo, J., Fakhrul Islam, S.M., Nurah, G., Khan, M.I., Staal, S.J. and Dugdill, B.T. 2004. *Employment Generation through Small-scale Dairy Marketing and Processing: Experiences from Kenya, Bangladesh and Ghana*. FAO Animal Production and Health Paper, no. 158. Rome: FAO.

Scholten, B.A. 1990. Europe's milk quotas ... six years down the road. *Hoard's Dairyman*, July, 587.

Scholten, B.A. 1998. *Dairy India 1997* (book review). *Food Policy*, 23(2), 211–12.

Scholten, B.A. 2010a. *India's White Revolution: Operation Flood, Food Aid and Development*. London, New York, New Delhi: I.B.Tauris, Palgrave Macmillan, Viva Books.

Scholten, B.A. 2010b. USDA organic pasture wars: shaping farmscapes in Washington State and beyond, in *Geographical Perspectives on Sustainable Rural Change*, edited by D.G. Winchell, R. Koster, D. Ramsey and G. Robinson. Brandon, MB: Rural Development Institute, 141–53.

Scholten, B.A and Basu, P. 2009. White counter-revolution? India's dairy cooperatives in a neoliberal era. *Human Geography*, 2(1).

Shah, M.K. and Dugdill, B.T. 1979. *Working Paper on a Proposal for Safeguarding the Development of the Domestic Dairy Industry in Bangladesh by Coordinating Local Milk Production with the Import of Dairy Products*. No: 080516, Project: Cooperative Dairy Organization, Bangladesh, BGD/73/003.

Singer, H. 1991. *Food Aid and Structural Adjustment in Sub-Saharan Africa*. Oxford: OUP.

World Bank. 2008. Can India's dairy revolution be repeated in Africa? World Bank [Online]. Available at: http://web.worldbank.org/WBSITE/EXTERNAL/NEWS/0,contentMDK:21936907~pagePK:34370~piPK:34424~theSitePK:4607,00.html [accessed: 2 March 2009].

World Bank. 2009. East Africa: World Bank approves US$90 million for agricultural productivity program. World Bank [Online]. Available at: http://web.worldbank.org/WBSITE/EXTERNAL/NEWS/0 [accessed: 14 July 2009].

Yunus, M. 2008. *Creating a World without Poverty: Social Business and the Future of Capitalism*. Jackson, TN: Public Affairs.

Chapter 12

Explicitly Licit: Stemming the Sand Tide in Kohsan District, Herat Province, Afghanistan

Lenard Milich[1]

Agriculture is robust in Afghanistan – but only for one specific crop. While there has been appreciable recovery in the agricultural sector overall since the fall of the Taliban regime in 2002, this has been made possible by enormous infusions of donor funds. Only this one specific crop showed production resilience during the past three decades of war that commenced with the Soviet Union's invasion of the country in 1979 – the opium poppy (Girardet 2011). Poppy has perhaps one-sixth of the water requirements of wheat, large areas can be sown with just a half-cup of seeds, it is relatively resilient to pests and disease, farmers have easy access to credit if they agree to cultivate it, and the return on investment (depending on the global market price) is usually several-fold more than licit crops (AREU n.d.).[2]

The United Nations Office on Drugs and Crime (UNODC 2010) estimates that during the past decade, some 70–90 per cent of the world's supply of illicit opiates came from Afghanistan. Revenues from this crop are equivalent to (depending on who is making the estimate) from approximately one-third of the licit agricultural sector GNP (UNODC 2003: 63) to as high as 35 per cent of total GNP (Glaze 2007: 5). There are clear linkages between opium poppy cultivation and the funding of the anti-government insurgency; the Taliban are known to tax production and each component of the trafficking pipeline within the country, but so too do pro-government 'warlords' (UNODC 2008: 32; Starkey 2008; Risen 2008; Goodhand 2008). The drugs issue, and its linkage with the omnipresent violent conflict, underpin the dangers of working in the country. External actors (donors, the UN, INGOs) face stringent security restrictions, and are often confined to secure compounds while Afghan nationals implement programmes and projects. Fortunately, ten years of capacity building across all sectors has by now greatly increased the capabilities of the Afghans themselves, but without a doubt, more remains to be achieved.

---

1   The views expressed herein are those of the author and do not necessarily reflect the views of the United Nations.

2   The Afghanistan Research and Evaluation Unit (AREU) has conducted a large volume of research on opium cultivation: see http://areu.org.af and search for "poppy."

A coalition of international donors has floated Afghanistan's public expenditures since the collapse of the Taliban regime in 2002. Figures from 2006–10 indicate that the international community ploughed $62 billion in civilian aid into the country on- and off-budget, with US assistance alone providing an estimated $38.5 billion (GAO 2011). Allegations abound regarding large-scale diversions of aid to Swiss bank accounts by the Afghans themselves, and to the personal bank accounts of international contractors and development workers.[3] According to SIGAR (2010) "[C]orruption, widely acknowledged to be a pervasive, systemic problem across Afghanistan, corrodes the Afghan government's legitimacy and undermines international development efforts". Accusations fly regarding the presumed benefits of development initiatives to the putative beneficiaries – the Afghan people – and it is undisputed that there are too many instances where the intended result has not proved to be durable beyond a project's closure, let alone sustainable (Arnoldy, 2010). The U.S. Senate Committee on Foreign Relations (2011) obliquely refers to this: "…lead to redundant and unsustainable donor projects, and fuel corruption." More explicitly, discussing the deluge of aid money disbursed in Helmand province, "[m]ost of the money was spent on unsustainable, short-term, cash-for-work programs that amount more to political handouts and buying love. Thus, two years later, they have little to show for themselves" (PBS, 2012). The billions in civilian aid, together with the assistance rendered by (military) Provincial Reconstruction Teams, should perhaps have had greater visible effect – after all, while there has never been a census in the country, at most the population may total 35 million. Questions arise as to why, after a decade of effort in blood and money, Afghanistan languishes at rank 158 of 172 countries in the 2010 Human Development Index, the lowest ranking in Asia.

With all this money disbursed, a large part of it designated to agricultural revitalization, a dispassionate observer might think that a system of rational triage would have focused on rehabilitation of primary canals, the lifeblood of most rural communities. Thus the UNODC, which led a 2009 Joint Assessment Mission (with eight other UN agencies) in the western Afghan provinces of Herat and Ghor, was surprised to discover that Kohsan District, on the border with the Islamic Republic of Iran, remained in dire straits. Eight years after the international coalition's development efforts had commenced, the Kohsan canal – lifeline to the 30,000–50,000 inhabitants of the district – was effectively out of commission due to Aeolian sand movement. True, the area is each year under the adverse summer wind conditions known as the Farah Bad, also called the 120-day wind (although nowadays local residents are aware that it lasts closer to 150 days, and that in winter a strong reverse wind blows too). Sustained wind speeds of 90 km/h are not uncommon during the summer months. Google Earth images of Kohsan show a veil of wind-borne sand over the area. But without the capacity to grow crops, what else would the population be doing to attain

---

3   There are many such allegations across a range of media outlets, but for an authoritative report, see for example SIGAR 2011.

household food security except resort to a suite of off-farm labour strategies that would include illicit ones associated with the opium economy?

There have been donor-funded efforts on the Kohsan canal system. The US Provincial Reconstruction Team had built an intake diversion in the Harirud River, but spring floods the following year had undermined half of it, collapsing the southern diversion wall. The World Bank had provided funds for rehabilitating the Ali Abad canal, a secondary canal in the district, but by late 2009 the concrete sluice was broken in several places. The *mirab bashi* (the water master over the whole canal system) was relatively unconcerned about these problems, however: his primary focus was on the inability of traditional community mechanisms (which in Afghanistan provide household labour or cash-in-lieu for canal cleaning based on the number of jeribs irrigated; one jerib is 2,000 square metres) to cope with the sheer volume of moving sand, which was in-filling the canal, smothering fields, inundating houses and causing respiratory ailments in the 14 villages of Kohsan district. Indeed, his concern and distress was based on clear evidence; UNODC's staff climbed artificial canal banks as high as 4 metres consisting of material that had been hand-dug from the main canal.

Mention of 'traditional community mechanisms' should trigger the question as to why these were disrupted in the first place. Was it simply a consequence of the years of drought earlier in the first decade of this century? That certainly played a role, but the marked degree of land degradation is a negative synergy of factors, the nexus between conflict and what appears to be the local result of global climate change, writ in the drylands of western Afghanistan.

As with any other locale in Afghanistan, Kohsan has often been enmeshed in conflict over the past 30 years. It all started in 1979, with the Soviet invasion. During the ten-year occupation, which Afghans refer to as the Mujahidin War, Kohsan district became the logistics gateway for the anti-Soviet forces for supplies from Iran across what was then an open border. The Soviets frequently staged attacks in the district, by both land and air. To protect their supply routes, and to cause as much havoc among the enemy as possible, the mujahidin sowed anti-tank minefields in several places in the district – a fact that has bearing on UNODC's activities today, as is explained below. During the decade of Soviet occupation, the consequence of Kohsan being a battle zone was that all orchards and household gardens were destroyed, the canal system was severely damaged, and most irrigated land (for the cereal crops of wheat and barley, as well as cotton, alfalfa, melon, watermelon and vegetables) put out of commission. This outcome was hardly unique to Kohsan. The Soviets practiced a scorched-earth policy in rural areas in an attempt to force the populace to move to the (more easily controlled) cities. "The effect was the complete collapse of the agricultural production of Afghanistan: the destruction of orchards, irrigation canals" (PBS, 2012). Prior to 1979, the district was renowned for its livestock, with available pasture and fodder resources for approximately 100,000 sheep and goats, many sold to Iranian markets; during the war, the livestock sector was badly damaged. Also prior to 1979, the wheat harvest was large enough that the considerable surplus could be sold to the government; as a result of the war, the entire agricultural economy was in tatters. Consequently, some 90 per cent of the populace was displaced, becoming internally displaced persons (IDPs) within Afghanistan or refugees outside, primarily in Iran because of its easy proximity.

After the Soviets departed in 1989, people gradually began to return to the devastated district, but agricultural infrastructure was so curtailed that household coping mechanisms had to shift. People started uprooting the desert bushes as household cooking fuel and for sale to Eslam Qala town (the main official border crossing point between Iran and Afghanistan, which lies in Kohsan district), or even as far as Herat city, 90 km distant. In desperation, households ploughed up the surrounding steppe, planting cumin and caraway. These did not yield well, but the disturbance of the soil surface, and the increasing harvest of bushes, is of course what triggered the Aeolian sand crisis now needing to be addressed.

The Taliban era saw no improvements in the agricultural situation. The historical record shows that in some places the Taliban replaced knowledgeable and experienced agronomists with mullahs who knew nothing about agriculture, and who could only quote from the Holy Qur'an whenever farmers approached them with problems.[4] The battles between the mujahidin forces and the Taliban displaced some small percentage of the population once more, but the main impact of the Taliban era was an abdication of government control over household agricultural practices, so that people cleared ever-larger tracts of land for cumin and caraway in a desperate attempt to stay ahead of declining yields suffered during a decade-long drought.[5]

Once the international coalition took over in 2002, a very few international NGOs returned to Kohsan, offering limited job opportunities, and agriculture and animal husbandry started on the slow road to recovery. Yet the sector languished, a prisoner to the moving sand problem and the enormous drop in the Q-factor, the canal's water-carrying capacity. Naturally, given the scarcity and unreliability of irrigation water, some people had turned to the cultivation of opium poppy, which yields well (and very lucratively) on around one-sixth of the water required for wheat.

Today, the situation remains serious: agriculture is still in the intensive care unit. Unlike neighbouring Iran, there are no curbs on population, no family planning programmes, and the population today is about four times as large as 30 years ago. Even with improvements in agricultural techniques and with external assistance, because of the total absence of rain-fed yields nowadays some 80 per cent of households must purchase wheat. The halcyon days of selling surplus to government are long past, the existence of an agricultural cornucopia a memory only of the communities' elders.

---

4   Personal communication with the pre-Taliban era head of a Provincial Directorate of Agriculture, triangulated with conversations with villagers.

5   Events over the past 30 years in Kohsan were elaborated to the author in personal communications with the *mirab bashi* and select community elders.

Yet by late 2009, when UNODC led the assessment mission, there were some positive signs that the inhabitants of the area recognized their role and responsibility in triggering the Aeolian sands. There was a blanket ban on the uprooting of bushes and on the grazing of livestock on the surrounding steppes (which by then had desertified), as well as on ploughing and the cultivation of cumin or caraway. The *mirab bashi* and the district governor had together designed what was on the verge of being a common property management regime. Violators of the bans were to be fined initially, and continued violations would result in jail time. But beyond this, one very positive autonomous development – plausible because of the proximity of the Iranian border – was the transition to cooking meals with LPG by almost all households, the large cylinders being transported back and forth across the border with Iran, the supplier. Finally, the canal's communities had banded together and purchased a used mechanical digger for dealing with the rapidly in-filling canal, albeit that purchase had exhausted their available funds. They had no money to purchase fuel or spare parts, or to pay the operator's salary. UNODC came to understand that the communities were primed for positive action. And so started UNODC's assistance, ultimately made possible by the people of Denmark, Finland and Japan. The first steps were obvious: to deal with the canal issue, to set up a land stabilization committee, and to find a reliable and experienced implementing partner to carry out the fieldwork, which proved to be the Afghan NGO DACAAR (Danish Committee for Aid to Afghan Refugees).

A couple of items are noteworthy before moving on to describe the work and its impact. First, the UNODC is a small agency in the United Nations, with scant resources to implement such work. In early 2010, at a donors' meeting at the agency's Vienna headquarters, one donor representative asked the author for the justification for implementing such a project (instead of, as the argument was made, focusing solely on building the capacity of the Ministry of Counter Narcotics to 'mainstream' alternative livelihoods, 'capacity-building' government counterparts being one of UNODC's principal global mandates). The answer was simple: because the need is there, because people are cultivating poppy and trafficking in it, and because after eight years, nobody else has done anything to address the problem of in-filling of the canal, the lifeline to the 14 villages that depend on it. But the question really does hit on target: UNODC is not in theory a 'line agency', so that in principle it is unquestionably true that implementing projects is outside its normal remit. Yet to a significant degree, other project implementers do not explicitly address the counter-narcotics issue; while Western donors may bemoan the cultivation of drugs, few are doing much themselves, or through their implementing partners, to explicitly address the conundrum of drugs cultivation. UNODC does; it fields a process that it labels 'genuine mainstreaming', making the counter-narcotics component central to the undertaking.

But it is also germane to point out that the UNODC has never explicitly endorsed eradication. The agency acknowledges the need for government to show its seriousness in tackling the drugs problem, and eradication is certainly one tool – if used appropriately and judiciously. The underlying problem, however, is that all too often promises made (whether by government, donors or others) for development assistance remain undelivered, while the demand for cessation of poppy (and cannabis) cultivation is made loud and clear. The carrot is missing, but the stick applied. Moreover, it is tragically the case that it is frequently the most powerless farmers, among the poorest in the community, whose fields are eradicated. Wealthy and well-connected farmers, and those who wield power from the barrel of a gun, are rarely if ever targeted (Draper 2011). Forcible eradication of the poppy fields of poor farmers could very well have the opposite result of that intended by forcing their households to obtain loans to buy daily rations; such credit is often extended by drugs middlemen, and usually the only way to pay it off is by growing poppy in the future.

Instead, UNODC uses a much softer, community-based approach, which it terms 'genuine mainstreaming'. This is notable in what it is not – it is not aid conditionality, in which we threaten to withhold assistance if the beneficiary community does not abide by the terms of agreement. Instead, we ask the *shura* (the council of elders) of each community we work with to endorse a 'social contract': that is, to pledge to eliminate opium poppy cultivation, cease trafficking in opium, heroin or heroin's precursor chemicals, advise its young men to avoid itinerant seasonal labour in the country's main poppy fields (in Helmand and Kandahar provinces), and to ask addicts (of which there are many in Afghanistan) to seek treatment. All enforcement is left in the hands of the shura, and explicitly noted in the agreement UNODC asks the shura members to sign is the request that they quote appropriately from the Holy Qur'an, reinforcing the idea that opium cultivation is haram (that is, forbidden). It then becomes a matter of honour for the shura to implement the agreement, and in the case of Kohsan it has been a successful strategy. A horrified shura approached a farmer who was growing opium poppy in between rows in his orchard, and he was asked to cease and desist – which he did.

A July 2011 internal assessment by DACAAR shows the following results pertinent to this counter-narcotics strategy:

Reduction in poppy cultivation: Poppy cultivation in the area has decreased by 93 per cent. In 2009, prior to the initial project assistance, 28 jeribs of land were under poppy cultivation in the 14 villages, decreasing to 2 jeribs in 2011.

Opium use among men and women decreased: Opium use among men and women has decreased by 30 per cent. Female community leaders report that since 2009, the number of opium addicts in the 14 villages has declined from 206 to 144 now.

Decrease in drugs trafficking through Kohsan district: In 2008–9, 30 people were arrested or killed in relation to opium trafficking within Kohsan district. However, there was not a single case of arrest or interdiction-triggered killing in connection with drugs trafficking in the period of 2010–11. It is indicative that such arrests and killings were still present in the neighbouring districts of Ghorian and Golran during the same period. According to village heads and mullahs, 20 per cent of households used to be involved in opium trafficking. They used vehicles, animals or simply human power to traffic opium across the border with Iran. According to these same sources, 5 per cent of households may still be involved in trafficking. The project has certainly contributed to the reduction of the trade (UNODC has interviewed one former trafficker noted as 'dangerous' by other members of the community, who today is a contented member of the Land Stabilization Committee), but other factors have also played a role. These include tightening of controls on the Afghan as well as Iranian side of the border, making it more difficult to enter Iran. Traffickers caught by Iranian authorities may be shot if resisting or evading arrest, and if found guilty during a trial are often executed. Moreover, the law in Afghanistan in relation to opium trafficking has become harsher and stricter. Community key informants are adamant that it is the social contract that each village shura has signed that has made the most important contribution.

Decrease in short-term labour migration to southern provinces of Afghanistan: There has been an 80 per cent decrease in people moving to Helmand and Kandahar as itinerant labourers for poppy cultivation and lancing (i.e., opium harvesting). During the 2009–10 opium harvesting season, 135 people left for these southern provinces to participate in lancing activities, according to the project coordination committee leaders. In 2010–11, only 26 people migrated for the same purpose.

However, it is likely the case that such gains are due to project-related short-term job creation, which provides an incentive to stay in the villages and work for wages. It raises the issue of durable gains, and sustainability of project activities. If return on investments for opium or cannabis is many times greater than for licit crops, the rational peasant model posits that illicit cultivation must necessarily continue, all else being equal. If off-farm labour opportunities fade away and are not replaced, no adult will complacently sit idly at home while his or her children starve. Hence, UNODC recognizes that its investment and involvement in Kohsan (and indeed in other locales in Afghanistan in which it is becoming involved) must be long-term. Decades-long, if necessary.

This too is, in general, what sets UNODC apart from many other development actors, who may be mandated to define and implement a fast exit strategy based more on political needs than on durable outcomes. UNODC has exit strategies for specific project components, but regards its partnership with the Kohsan communities as being open-ended, reaching into the indefinite future. For example, the most recent initiative started will, if successful, transform the Land Stabilization Committee into a Pasture Users' Committee, modelled on the new common property resource management regime developed in Kyrgyzstan under its 2009 law On Pastures. In essence, such localized pasture management supplants quantity of livestock with livestock and pasture quality, assuring a sustainable use of the pasture resource and optimizing household income streams. Since pasture vegetation will not recover for another two to three years in optimal circumstances, it may be five years before a Pasture Users' Committee will commence operational control over pastures. But the preparatory work – sensitization of communities, training in pasture management and vernacular Monitoring and Evaluation (M&E) procedures for pasture quality monitoring, establishing standard procedures for irrigated fodder crops, involvement of para-veterinarians in proselytizing the benefits of preventative care – can, and indeed should, begin now.

But that is the future. A review of past and present initiatives will show how UNODC in fact edged into acknowledging the need for long-term commitments as a lesson learned, rather than having a comprehensive set of complementary and synergistic projects determined in advance. This is, of course, logically likely to be the model in any post-conflict effort to rehabilitate agricultural infrastructure and revive productivity – address first the emergency, then move on to learning what can be done to boost household livelihoods. It is a piecemeal approach, but one that may actually work out better in the long run because it involves the building of trust, of a recognition by beneficiary communities that their lives are gradually improving – and that the lives of their children could be substantially better. Psychologically, such an approach is likely to yield dividends over the years of activities that a single large infusion of assistance cannot.

The emergency response took place in January and February 2010, when UNODC rented a mechanical digger (with operator and fuel) and two trucks used to ferry 200 community labourers back and forth along the Kohsan canal's 54 km length. UNODC partnered with the UN World Food Programme, which provided food-for-work incentives for the labourers. This initiative was led in its entirety by the *mirab bashi*, and UNODC made certain that it boosted this man's profile as much as feasible, since he is a critical actor and agent for change. For example, UNODC invited him to attend a planning meeting at the district governor's office, then delivered him back to his village home in a UN vehicle, making sure that his neighbours witnessed the respectful gesture.

Of course, addressing just the symptom of sand blocking canal flow would have been an exercise in futility. Thus in the absence of anyone's capability to diminish the wind's velocity, the next obvious step was to ascertain how to stabilize the land surface. Regionally, soil surface ablation (by wind, or rainstorm sheet and/or splash erosion) is not an uncommon problem, and Iran has had procedures in place for decades to stop (or at least retard) the losses. But UNODC wanted to utilize a local NGO if possible, and was pleased to learn of DACAAR's previous land stabilization efforts in another district of Herat province – the first ever in Afghanistan. Initial contacts led to a full project proposal, and so began a successful partnership – successful in two specific regards: first, UNODC's initial agreement with the *mirab bashi* was that the digger would be provided for two consecutive years, but he informed us that this year (2011) this assistance would not be required, since traditional community mechanisms could now cope with the routine canal maintenance; second, Denmark is prepared to support a roll-out of the work in other parts of western Afghanistan similarly affected by blowing sand.

DACAAR's role in land stabilization commenced shortly after the canal cleaning ended in early 2010. It implemented a participatory approach, forming a committee that included village elders from the 14 villages. Terms of Partnership were agreed among the committee members to define the responsibilities of all stakeholders, followed by the development of Terms of Maintenance (ToM). Agreed by the communities, ToMs included detailed action plans for project implementation. The newly formed Land Stabilization Committee was expected to meet every six months to discuss progress and challenges. A 10 per cent contribution from the community in the form of labour was agreed in order to increase their ownership of the project. A total of 6,034 people received soil stabilization messages, of which 40 per cent were women. They learned of the importance of vegetation regeneration, the need for maintenance of the project area and for community contribution, and the dangers of opium production and use.

The land stabilization project initially focused on 4,000 hectares most at risk from soil erosion, with the aim of placing this area under protection through augmentation of vegetation cover. It was to include community participation, use of appropriate technologies and the planting of suitable species. The initiation study concluded that soil stabilization needed continued support for at least another three years to ensure sustainability of community support and sufficient change in vegetation to decrease soil ablation. With Danish and Japanese support, in 2011 the pilot project was expanded to protect 10,000 ha of agricultural land, canals and dwellings from sand encroachment through the natural regeneration of native vegetation, together with the introduction of drought-resistant plants and eyebrow water-harvesting techniques. The project further aimed at improving livelihood conditions of target beneficiaries through short-term job creation and the building of entrepreneurial capacities. In late 2010, UNODC/DACAAR commissioned consultants from the University of Mashhad, Iran, to advise on the introduction of new crops, basing their analysis on comparable climate and edaphic factors between Kohsan and areas in Iran. (As a by-product, UNODC introduced these Iranian researchers to NASA's Giovanni/TOVAS system, which generates online analysis of spatial rainfall at $0.25°$ resolution.) Thus far, new watermelon varieties, almond and saffron have been introduced.

With the aim of bringing 10,000 ha of land under protection in 14 villages of Kohsan district, by 2010, 6,000 ha had been brought under the protection scheme. By May 2011, 2,000 further hectares had been added, bringing the total to 8,000 ha of protected land. By the end of 2012, this will reach the 10,000 ha. In the protected area, drought-resistant bushes (*Atriplex* and *Haloxylon*) have been introduced. By the end of 2010, 40,000 Haloxylon seedlings were planted in the area, with an average survival rate of 63 per cent, as well as 26,000 Atriplex seedlings imported from Iran having even lower survival rates – just 10 per cent (an issue explored further below). Additionally, 15,000 ash and mulberry trees have been planted as a windbreak along the north bank of the Kohsan canal, with high survival rates for ash but low for mulberry.

Results to date are encouraging. These include:

Environmental – In the target village areas, degradation of vegetation cover has reversed, showing positive from its pre-project negative trend; the percentage of plant cover has improved from 14 per cent to 24 per cent and plant litter has increased by 8 per cent. The communities of target villages contend that there has been 40 per cent decrease in moving sand and dust. Key informants report that now there is less sand inside houses and it's possible to keep windows and doors open. The communities have demonstrated commitment toward the continued protection of target areas; they hired a local guard to exclude grazing and prevent the harvesting of biomass.

Improved health and hygiene – There has been about a 40 per cent reduction in respiratory diseases and eye problems in 2011 as compared to 2009, prior to the project's start. One village elder mentioned that before the project commenced, children had to wear glasses to prevent sand entering their eyes, and were not enthusiastic about leaving home to go to school. However, this situation has improved considerably. Women's Community Development Council leaders reported that drinking water and hygiene has improved due to less sand in food and drinking water.

Agriculture yield increase – Key informants state that they expect agricultural yield to increase by 35 per cent this year. Three factors (all attributable to the project) contribute to this result: (a) there was less erosion of soil by wind; (b) the land under cultivation increased by 35 per cent; and (c) there was 48 per cent more water available for irrigation.

Reversal of out-migration trend – It was until recently quite common for entire households to migrate to other provinces to escape the deteriorating environmental and socio-economic situation of Kohsan. This trend has now reversed – 136 families left Kohsan in 2009 to settle somewhere else; in 2010–11, not a single family left Kohsan, but 154 families either returned or are new in-migrants, former IDPs. Of course, there may be other factors at play in the increasing trend of reverse migration in Herat province. The Iranian government is becoming much stricter in regard to illegal Afghan workers in Iran, with daily deportations of Afghans. In 2008–9, 599 individuals left from Kohsan for Iran to work; in 2010–11, the figure stands at only 92.

Gender mainstreaming – Progress is being made in building entrepreneurial skills among targeted women. In 2010, 200 women started Atriplex nurseries in their kitchen gardens. They produced 26,000 Atriplex and sold them to DACAAR for 8 Afs (USD$0.19) per seedling. In 2011, another 350 women started Atriplex nurseries. The project-sponsored Women's Resource Center has developed two business plans, with two more to be elaborated in coming months. The project is expected to benefit the most vulnerable women in these villages through alternative crops such as saffron, as well as improved animal husbandry (in Afghanistan it is mostly the women who take care of small ruminants). Women feel that if they can increase household incomes through animal husbandry and alternative crops, the men in their families will not have to be involved in illicit activities, or migrate to other provinces or Iran for work.

There have been significant hurdles to overcome, as the low seedling survival rates for 2010 illustrate, and lessons that needed to be learned. Haloxylon seedlings initially had to be imported from Iran in 2010, but due to bureaucracy and logistics, planting was delayed and the project faced problems with available labour for irrigation. The result was low survival rate of the saplings (63 per cent). For 2011, seeds were planted early in the year and the seedlings could be transplanted on time, resulting in a 95 per cent survival rate. A similar situation existed for the mulberry saplings; imported from Iran, they were too large when dug up for transplanting, resulting in the severing of roots, the desiccation of root balls in transit, and low survival rates. This year, smaller saplings with intact root systems will be planted to replace the dead mulberry saplings. Atriplex faces more difficulty: it thrives on saline, loamy soils, but most of Kohsan consists of sandy soils; further, the strong winds stripped the leaves from the Atriplex seedlings, without which photosynthesis becomes impossible and the plant dies. Nonetheless, in more sheltered, edaphically suitable areas, 80,000 Atriplex seedlings will be planted in early autumn 2011 to take advantage of seasonal rains.

The presence of the anti-tank minefields has had a negative influence on the project. While most minefields in the country have been mapped, there were several in the project area that had not been included. The work plan initially called for irrigation of seedlings to be carried out from a water tanker drawn by a tractor, but when the residents balked at the idea, the presence of the minefields came to light. Now, most irrigation is done by hand, with buckets filled from the tanker by labourers who walk across the desert. While in general terms this amounts to a lesson learned, in fact there is nothing that could have been done differently as the land stabilization effort has to focus on friable source regions and emplace windbreaks where required, not where convenient. Foreknowledge of the presence of minefields would not have generated a different project pathway, since mine removal would multiply project costs several-fold, requiring expenditures approaching US$3 million.

Finally, many parts of western Afghanistan are eminently suitable for the cultivation of saffron, a new crop that DACAAR first introduced into the country in 1998. Recently, saffron has been much touted as the counter-narcotics, anti-poverty 'magic bullet', due to its (usually) high sales price on global markets. While some saffron corms originated from the Netherlands, they were not suited to harsher Afghan growing environments. Most of Afghanistan's saffron industry originates from Iranian corms. But the catch is that Iran categorizes saffron as a 'strategic crop', the outcome of that labelling being that corms from Iran had to be smuggled across the border into Afghanistan. But, as UNODC pointed out to many others during 2011, Iran was unlikely to sit idly by, simply watching Afghanistan's increasing competition. And indeed, the global price for saffron has plummeted this year, providing stringent disincentives to Afghan producers because of the immense labour costs required to harvest and process the spice. At the time of writing, Iranian saffron at ISO standards is available in Kabul at retail prices lower than Afghan saffron, which due to ineffectual quality control procedures regularly fails tests for bacterial and/or aflatoxin contamination. This market situation is likely to prevail for some years. The project has now acquired the list of 'strategic crops' from Iranian authorities, and will expand its introduction of new crops while avoiding Iranian ire.

UNODC has also implemented several complementary activities in Kohsan; with the United Nations Assistance Mission in Afghanistan (UNAMA), it conducted M&E training for district government officials; with the World Health Organization, it offered basic health worker training to 75 men and women; with the University of Herat's Faculty of Agriculture, it provided training of trainers in agricultural innovation to key farmers; and it sent its own staff to the Symposium on Central Asian Pastoralism in Kyrgyzstan, leading to the aforementioned initiative for the creation of a Pasture Users' Association and its governing Committee.

The metanarrative lessons the author and the organization for which he works have learned include the desirability to design a series of interconnected activities that allow reaffirmation of commitment to the beneficiary communities over time; to 'stay the course' if sustainability is to be achieved, aiming for an ultimate tipping point that will enable the communities to thrive on their own; to be flexible enough to add innovations at any time; to not presume that anyone involved in the work knows enough about all possible ways forward, and therefore to read widely in the search for new keystone ideas that can be immediately applicable and effective; to do the basic research into proposed new crops so as not to be surprised by changing markets or social conditions that can cause initiatives to collapse; and finally, to never cease to be inspired by the people we serve.

Acronyms

DACAAR	Danish Committee for Aid to Afghan Refugees
M&E	Monitoring and Evaluation
UNAMA	United Nations Assistance Mission in Afghanistan
UNODC	United Nations Office on Drugs and Crime

References

AREU. *'Poppy'* [Online]. Available at: http://areu.org.af [accessed: November 2011]

Arnoldy, B. 2010. 'Afghanistan war: How USAID loses hearts and minds'. The Christian Science Monitor. Available at: http://www.csmonitor.com/World/Asia-South-Central/2010/0728/Afghanistan-war-How-USAID-loses-hearts-and-minds [accessed: August 2011].

Draper, R. 2011. Opium wars. *National Geographic* [Online]. Available at: http://ngm.nationalgeographic.com/2011/02/opium-wars/draper-text/1 [accessed: August 2011].

GAO 2011. Afghanistan's Donor Dependence [Online]. Available at: http://www.gao.gov/products/GAO-11-948R [accessed: February 2012].

Girardet, E., 2011. *Killing the Cranes: A Reporter's Journey through Three Decades of War in Afghanistan.* White River Junction, VT: Chelsea Green Publishing.

Glaze, J.A. 2007. *Opium and Afghanistan: Reassessing U.S. Counternarcotics Strategy.* Strategic Studies Institute, United States Army War College [Online]. Available at: http://www.strategicstudiesinstitute.army.mil/pdffiles/pub804. pdf [accessed: August 2011].

Goodhand, J. 2008. Poppy, Politics, and State Building. In Hayes G. and M. Sedra (eds.) *Afghanistan: Transition under Threat.* Wilfred Laurier University Press, 51–88 [Online]. Available online at: http://dspace.cigilibrary.org/jspui/ bitstream/123456789/27397/1/Afghanistan%20Transition%20under%20 Threat.pdf?1 [accessed: February 2012].

PBS, 2012. *Why Eradication Won't Solve Afghansitan's Poppy Problem.* Interview with Dr. Felbab-Brown of the Brookings Institution [Online]. Available at: http://www.pbs.org/wgbh/pages/frontline/afghanistan-pakistan/opium-brides/ why-eradication-wont-solve-afghanistans-poppy-problem/ [accessed: February 2012].

Risen, J. 2008. Reports link Karzai's brother to Afghanistan heroin trade. *New York Times*, 4 October [Online]. Available at: http://www.nytimes.com/2008/10/05/ world/asia/05afghan.html/ [accessed: August 2011].

SIGAR 2010. *U.S. Reconstruction Efforts in Afghanistan Would Benefit from a Finalized Comprehensive U.S. Anti-Corruption Strategy.* Office of the Special Inspector General for Afghanistan Reconstruction [Online]. Available at: http://www.sigar.mil/pdf/audits/SIGARAudit-10-15a.pdf [accessed: February 2012].

SIGAR 2011. *Limited Interagency Coordination and Insufficient Controls over U.S. Funds in Afghanistan Hamper U.S. Efforts to Develop the Afghan Financial Sector and Safeguard U.S. Cash.* Office of the Special Inspector General for Afghanistan Reconstruction [Online]. Available at: http://www. sigar.mil/pdf/audits/SIGAR%20Audit-11-13.pdf [accessed: February 2012].

Starkey, J. 2008. Drugs for guns: how the Afghan heroin trade is fuelling the Taliban insurgency. *Independent*, 29 April [Online]. Available at: http://www. independent.co.uk/news/world/asia/drugs-for-guns-how-the-afghan-heroin- trade-is-fuelling-the-taliban-insurgency-817230.html [accessed: August 2011].

United Nations Office on Drugs and Crime (UNODC). 2003. *The Opium Economy in Afghanistan: An International Problem.* UNODC [Online]. Available at: http:// www.unodc.org/pdf/publications/afg_opium_economy_2003.pdf   [accessed: August 2011].

United Nations Office on Drugs and Crime (UNODC). 2008. *Illicit Drug Trends in Afghanistan.* UNODC [Online]. Available at: http://www.unodc. org/documents/regional/central-asia/Illicit%20Drug%20Trends%20Report_ Afg%2013%20June%202008.pdf [accessed: August 2011].

United Nations Office on Drugs and Crime (UNODC). 2010. *Afghanistan Opium Survey 2010.* UNODC [Online]. Available at: http://www.unodc.org/

documents/crop-monitoring/Afghanistan/Afghanistan_Opium_Survey_2010_
web.pdf [accessed: August 2011].

U.S. Senate Committee on Foreign Relations (2011). *Evaluating U.S. Foreign
Assistance to Afghanistan* [Online]. Available at: http://www.foreign.senate.
gov/imo/media/doc/SPRT%20112-21.pdf [accessed: February 2012].

Chapter 13

Olive Trees: Livelihoods and Resistance

Marwan Darweish

Introduction

Palestine enjoys diverse climatic conditions from the Jordan valley desert to the south and east to the mountainous areas and fertile planes in the north and west in which a variety of crops flourish. In the last few decades agriculture has proven to be critical for the fragile political situation in the Occupied Palestinian Territories (OPT) where unemployment, poverty and lack of food security have become a priority to the Palestinian National Authority (PNA). Since the Israeli occupation of the West Bank and Gaza Strip in 1967, Israel has used the Palestinian population as a reservoir of cheap labour for its own economy. Israel controls the Palestinian economy through direct and indirect measures so the PNA lacks independence and sovereignty and consequently is largely dependent on international aid to function and maintain its development and emergency programmes. According to the Palestinian Ministry of Finance (MOF) this assistance is estimated at \$11.7 billion from 1994 to 2010.

Agriculture and olive growing is critical for Palestinian livelihoods which have been based on a subsistence family economy for generations where the majority of agriculture and livestock remain for family consumption. It is estimated that Palestinian agricultural land is covered with some 10 million olive trees organized in rows on the terraces and that some 900,000 dunums, approximately 45 per cent of the arable land in the OPT, are planted with olive trees and account for the majority of fruit trees area.

The OPT is divided into three areas; A and B are under Palestinian civil control and area C under Israeli military and civil control, C is where 62.9 per cent of agricultural land is located so the presence of the Israeli military pose a challenge to farmers. In the Gaza Strip however, 17 per cent of the area has been designated by Israel as a security belt adjacent to the border and this has been increased since Israel's military offensive into the Gaza Strip in late 2008. Palestinians face travel restrictions within the OPT and entering and existing the OPT as Israel has imposed hundreds of checkpoints to control the movement of people and goods. This causes long delays and disrupts the daily life of the population. The difficulty in transporting goods to the cities has had an immediate impact on the income of families and their standard of living (UNDP 2011).

This chapter will explore the economic and cultural significance of olive growing for Palestinians in the OPT in the midst of violent conflict between Israel and the Palestinians. It will argue that the impact of the Palestinian Israeli conflict on the agriculture sector and olive growing specifically has been damaging. Palestinian farmers relying on the income from the olive harvest struggle to support the household. This chapter will provide a brief discussion of the direct and systemic structural violence and intimidation exercised by the Israeli settlers and army towards the Palestinian farmers such as land confiscation, access to land, injury and death to farmers and damage to their property and crop. The chapter will also highlight the contribution of the olive production sector to the resilience and resistance of the Palestinians to stay on their land, and the obstacles facing farmers in sustaining this sector. The chapter will explore the social and cultural features of olive harvesting and the role it plays in providing cohesion for the society. Finally the chapter will demonstrate the economic contribution of the olive harvest to the economy and resilience of the community under occupation.

No Peace for the Olive Branch

Israel has instituted a dual legal system in the West Bank; one for the Jews and another for the Palestinians. Jewish settlers have freedom of movement, political participation, infrastructure and access to water and land while the Palestinian population is denied its basic civil and human rights. The existence of settlements is a violation of international humanitarian law and causes grave and continuing infringement of the rights of the Palestinians. Since the occupation of the Palestinian territories in 1967, Israel has built 132 Jewish settlements in the West Bank in addition to many other unrecognized settlements or 'outposts' as Israel labels them.

Based on a complex legal bureaucratic system created by Israel, some of which was inherited from pre 1967, Israel has confiscated about half of the West Bank land for settlement building. This land was largely privately owned and suitable for farming (Darweish 2010). Physical barriers, fences and roads were built around the settlements but gradually removed to allow further expansion of the settlements. These are different from the barriers that prevent Palestinian farmers from accessing their land. Over the years Palestinian farmers working or using their land adjacent to the settlements have faced harassment, destruction of their property and violent attacks. Palestinian farmers often do not dare enter their land and risk their lives and the lives of the family members. Sometimes farmers only wanting to cross the land must be accompanied by international volunteers or Israeli human rights activists to reduce the chance or severity of attack. One Palestinian farmer gave testimony to B'Tselem, an Israeli human rights organization, 'ever since these settlements were built on our lands, we can't even get to the lands close to them. Every time we come near settlers they attack us..when we approach, they draw their weapons and aim them at us. They throw stones at us and beat us with clubs, all in front of the eyes of the soldiers guarding the settlements.' (B'Tselem 2008: 23)

Settlers' violence and harassment towards Palestinian farmers has been on the increase since 2000, the start of the second Intifada, to prevent them from accessing their land and consequently from cultivating it. Israeli and Palestinian human rights organization such the Palestinian Rights Monitor, Al-Haq, B'Tselem and Rabbis for Human Rights, and international organizations such as Human Rights Watch have documented various forms of violence and intimidation directed at the Palestinians including shooting at farmers, threats to shoot and kill, stone throwing, destruction of farms, property and equipment, theft and damage of crops, committing arson in the olive groves and more. This is part of a systematic policy of harassment and violence aimed at expelling Palestinians from land adjacent to the settlements, deny them access and to control the land as part of the creeping annexation of Palestinian land all with the silent consent of the Israeli authority. In 2006, heads of some Palestinian local authorities in the West Bank appealed to the Israeli Supreme Court to enable them to harvest their olive trees adjacent to the settlements. The Court held the state of Israel responsible and said that it must take all means necessary to ensure the safety of Palestinian farmers in these areas. 'Protection of the Palestinians must be done in a suitable manner, clear directives must be given to military forces and the police as to how to act, and effective restrictions must be placed on persons who harass Palestinians in breach of the law'. However, the challenge would be the implementation of such a decision and the impact of this directive has been sadly lacking (B'Tselem 2008: 25).

The argument put forward by the Israeli state for the confiscation of Palestinian land and denial of Palestinian access to their land is both expansionist and security driven; Israel argues that the land is needed for future settlement growth and to make the Israeli army's job of protecting settlers easier. Israel's policy includes loss of access to the land on which the settlements themselves were built and land west of the Separation Wall, roads on which only Israelis are allowed to travel, and lands for military use. The Israeli army is required to impose conditions and regulations on the Palestinian farmers' cultivation of their own land adjacent to the settlements and the fence surrounding it or the land west of the Separation Wall. The conditions include a) the recognition of ownership of the land by Palestinians, b) obtaining a set date for entry dictated by the army and c) consent by settlers to enter the land. Setting such conditions precisely reflects the dual legal systems in the West Bank as identified above. Evidently it is very difficult to meet the requirements for Palestinians to enter their closed off land and for this reason they have increasingly ceased trying to access their land. This causes frustration and hopelessness in the face of the bureaucracy set up by Israel. A Palestinian farmer affected by the Separation Wall explained the impact of the regulations: 'Three months with a permit, three months legal struggle to have to renew again, with prohibitive legal fees of 2000 NIS'. The farmer was told he was not given a permit by the Israeli Civil Administration for reasons of security. His three sons have never been given a permit to work on their land despite the fact that one of them has a permit to travel to Israel for business' (Parry 2004).

During the second Intifada the army accused Palestinians militants of using orchards adjacent to the Jewish settlement, and owned by Palestinians, as a shield and hiding place to attack the army and infiltrate the settlements and attack residents. The Israeli Army claim that as a 'defensive measure' it uprooted thousands of olive trees and destroyed cultivated land to increase visibility for soldiers and protect soldiers and settlers. Orders were issued by the army to increase the already enclosed land around the settlements, these became known as 'Special Security Areas' (SSA), covering a radius of about 300–400 meters of land. In 2004, Israel authorized plans and a budget to ring 41 settlements with SSA. As a result thousands of acres of land have been confiscated and trees uprooted and damaged.

The Israeli army held Palestinian farmers responsible for attacks on Israeli settlers for allowing Palestinian militants to use their olive orchards as hiding places from which to launch attacks. An Israeli Commander in the OPT made the argument that farmers are collaborating with the militants to use their land. He explained that 'owners of groves are to blame when their trees are uprooted. If the owner of the grove, whom I assume knows the sniper or the petrol throwers, does not take the measures he must take, then his grove will come down' (Sarafa 2004: 13).

Through this set of complex security measures and regulations imposed on the Palestinian farmers they have been systematically prohibited from free access to their land and as such any benefit from their olive orchards. This policy aims slowly but surely to deny them their rights on the land and make them powerless and dependent; they appear stripped of aspiration and identity and are more impoverished day by day. Despite these obstacles, olives and the olive harvest still play a significant role in Palestinian culture and tradition.

Social Cultural Aspects of Olive Harvest

Olive trees were brought to Palestine from Greece thousands of years ago and today they can be found on almost every hill and mount. Olive trees and the olive harvest have become part of the Palestinian culture and value system and are reflected in religion. Olive trees have a religious significance for all three monotheistic religions Judaism, Christianity and Islam, as a symbol of peace and forgiveness. The three religions share the story of Noah and the Ark where after the devastation of the flood a dove from the Ark brought an olive branch to Noah as a sign of life and appeasement. Olive trees are mentioned in all three holy books – The Torah, The Bible, The Quran – praising the qualities of olives and the land they grow on (ARIJ 2007).

It is a long tradition that families gather together to bring in the olive harvest from mid September for a period of up to two months. During the harvest season all the family members join together to collect the olives. Members of the wider family; men, women, young and old all participate. There is an archetypal sense of 'Thanksgiving' that peasant and farming communities throughout the world traditionally experience harvesting a valued and treasured crop. There is a feeling of togetherness where all the members of the extended family may be under the same olive tree for hours each day over a few weeks, from the morning until the evening and everyone in the olive orchard is focused on one objective.

There is an atmosphere of celebration during the harvest season. Usually traditional Palestinian folk songs and contemporary political national songs will be recited during the harvest and all the family members turn into an orchestra of production and wellbeing. Festivals and public celebrations are organized by local councils and farmers cooperatives and unions. More recently these celebrations have begun to reflect political resistance against the construction of the Separation Wall and the confiscation of Palestinian land, as an activist from Stop the Wall Campaign (SWC) explained 'we supported a few olive festivals in the area of Nablus, in 2009, to celebrate the end of the harvest season and encourage farmers' steadfastness in the face of attacks of Jewish settlers.' (SWC 2010)

The annual olive harvesting is an integral part of the tradition and life of the Palestinian family. It embraces different rituals during the harvesting and olives, oil, and 'zaater'[1] are part of the daily breakfast and food generally. Uprooting trees and preventing Palestinians from entering their land has not only caused economic hardship but has also had a psychological impact; this has caused untold damage to the way of life of the community as a whole. In a testimony to B'Tselem a Palestinian famer described that 'people not only bring forth bread from the earth, they also bring forth rest and relaxation..[it is] two years since school children have taken a hike in the bosom of nature, because the bosom of nature has disappeared beyond the fence, where they cannot go' (2009: 66).

The gathering of all members of the family everyday over a few weeks provides an opportunity to air views about social and political conflicts. Palestinian society is highly politicized and diverse in its views and one family can represent a spectrum of political views sparking in-depth debates and improving understanding over this time together. This is an informal process of consultation and listening to other views in the family to resolve conflicts constructively. It is a consultative and inclusive process conducted under the olive trees. Naturally disputes about land ownership among Palestinians surface during the harvest season and this provides an opportunity to have a dialogue between the disputant parties to resolve it peacefully.

One of the main root causes of the conflict between Palestinians and Israelis is over land. Olive trees and trees in general symbolize the conflict and connection of Palestinians to the land. They symbolize the rootedness in the land and the right of the Palestinians to be in their home land. Olive trees are 'seen as embodying the qualities of rootedness and durability, attributes Palestinians say they believed it preserved them during years of struggle with Israel.'(Lynfield 2000)

Olive trees have become the microcosm of the Palestinian Israeli conflict and vividly illustrate the destruction and damage resulting from the conflict. The plight of the olive trees can be seen to embody the conflict with the Israeli occupation and the

---

1   Middle Eastern herb (Thyme). This condiment made from the dried herb and mixed with toasted sesame seeds and salt.

Palestinian population; they are a symbol of the national struggle for liberation and resistance of the Israeli occupation. Political groups and civil society organizations arrange annual campaigns for trees plantations, harvest and work on the land to support farmers. These actions are seen as a way of expressing belonging to the land and resistance to the Israeli policy of land confiscation. Through plantation and agriculture work the community asserts a clear political position of resistance and steadfastness. A Palestinian farmer from inside Israel explained: 'during the seventies our land was under the threat of confiscation by the Israeli authorities. The land had been used for military training by the Israeli army without our permission. Planting olive trees on the land was an important act to prove our ownership and belonging to the land and to prevent the military training.' [2]

However in contrast to oppression and control, olive trees are also known to symbolize peace and tranquillity and as way of seeking peace as one may say 'to hold out an olive branch'. Olive branches have been offered as a symbol of peace for thousands of years due to the fact that it takes many years for olive trees to mature and produce fruit and can only be cultivated during long enough periods of stability and peace. After the signing of the Oslo peace agreement between Israel and the PLO in 1993, many Palestinians went into the streets to celebrate and as a gesture of peace they handed olive branches to the Israeli soldiers.

International and Israeli peace organizations express solidarity and support to Palestinian farmers in their struggle to resist the uprooting of olive trees and the harassment they face by settlers and the Israeli army. The Young Men's Christian Association in Palestine organizes international annual olive tree planting to support farmers. Every year volunteers from all over the world join farmers during the harvest season to help and to provide protection from the army and settler attacks. They plant and sponsor olive trees and work on the land as a form of solidarity and support. Rabbis for Human Rights, an Israeli peace organization, also supports Palestinian farmers when their trees are uprooted by the army or the settlers through purchasing new olive trees and replanting them as an act of solidarity and humanitarian assistance. The aim of the Olive Trees for Peace Campaign project organized by Rabbis for Human Rights is to enhance communication between Palestinians and Israelis and advocate for peace between the two peoples (Cheri 2001: 96).

Olive trees and the olive harvest reflect the political and religious symbolism in Palestinian culture and contribute to its cohesion and resilience. However, this culture and social significance has been threatened by the violence of the settlers and been turned from joy and celebration to fear and intimidation.

Olive Harvest: Fear not Joy

The relationship between the Palestinians and the settlers has been strenuous and frequently openly hostile since the occupation of the West Bank and Gaza Strip in

---

2   Interview conducted by the author, 2010.

1967 and the confiscation of Palestinian land to build Jewish settlements. Direct physical attacks on Palestinian olive harvesters by settlers are only one aspect of a broader system of violence and intimidation. During the few weeks of the olive harvest the level of violence significantly increases against Palestinians farmers and their families. During this period thousands of Palestinian families take part in the olive harvest which sometimes takes place on land adjacent to settlements or on Palestinian farm land traversed by settlers' roads. The conflict between the farmers and the settlers is manifested starkly over the land, the trees, the crop and the harvest (Oxfam 2010).

The olive harvest season is traditionally associated with joy and celebration where all the family comes together has taken on a darker significance. The harassment and constant violent attacks by settlers and the army have changed this atmosphere and made it a more frightening and traumatic experience. Farmers feel threatened by the presence of soldiers on their land despite their supposed role as protectors from settlers' attacks. These are the same soldiers responsible for the oppression and violence that the Palestinian face every day, that harass and hold them at the checkpoints in their daily life and have confronted them in years of conflict. One woman described this situation as follows: 'imagine that we pick the olives with the soldiers' right next to us, how can you feel safe? What can we do if we need to pray or go to the bathroom? We go where they can't see us and relieve ourselves under an oak or an olive tree. We do not feel safe and we are scared all the time. We have even stopped making tea and coffee over on open fire, and we bring it prepared from home. We eat fast to get back to picking and finish quickly' (B'Tselem 2008: 75).

Restrictions and conditions have been set up to prohibit Palestinian landowners obtaining permits to enter and work in their land. They are in a particularly vulnerable position and exposed to extortion by the Israeli security service to gather information about political and military activities amongst Palestinian groups. The Israeli security services exploit this situation to recruit Palestinian collaborators in return for providing entry permits for farmers. As one of the farmer described on Wednesday 8 August 2007 he was supposed to enter the land when he got a call from a civil administration official asking about gunshots in the village. The official asked the farmer for the names of the children who burned the fence and threatened the farmer that if he does not cooperate he will not have a permit to enter the land. The farmer believed that the official wanted to turn him into a spy (B'Tselem 2008: 75).

Uprooting and destroying Palestinian olive trees has been carried out by the Israeli army and Jewish settlers who take the law into their hands. It is well documented that settlers go into Palestinian orchards and vandalize olive trees and set them on fire. According to the Israeli human rights organization who presented photos of the vandalized trees to an expert on olive tree cultivation from the Hebrew University, 'the trees as young as the ones in the photographs would not normally be pruned, and they had therefore evidently been vandalized..and the damaged trees would take eight to twelve years to recover fully and produce as much as they had before.' A Haaretz journalist commented on this situation saying that 'there is something very human about these stumps of olive trees, hundreds upon hundreds of them, amputated branches reaching skyward as if to ask for help' (Hass 2006).

The Palestinians so far have had their land and olive groves confiscated, been intimidated when they can access it, experienced the destruction of their orchards but the olive groves have also been victims of the Israeli security laws. The construction of the Separation Wall between the West Bank and Israel has caused momentous damage to the Palestinian civilian population and also to farmers. Thousands of olive and fruit trees were uprooted and thousands of acres have been confiscated to make way for the construction of the Wall. The building of watch towers along the Wall and patrol routes has added to further confiscation of land and destruction and uprooting of olive trees. In Qifeen village in the North of the West Bank, for example, 12,600 olive trees were uprooted and more than 100,000 trees 'imprisoned' behind the Separation Wall. Farmers find it extremely difficult to cross to the west side of the Wall to cultivate their land given the restrictions and obstacles imposed by the Israeli army to obtain permits. Uprooting, destroying or confiscating land and trees has had significant effects on different aspects of Palestinian life. According to conservative estimates of Israeli human rights organizations, from March to December 2005 a total 2,616 trees were uprooted, burned, chopped or stolen (Hass 2006).

The impact of the systemic violence and intimidation is far reaching and profound on the livelihood and aspiration of the Palestinian population and in spite of this threats and fear, Palestinians continue to work in their land and enjoy harvest season and profit from the olives.

Economic Fruits

The contribution of the agricultural sector to the national Palestinian economy varies every year, however there has been a gradual increase since 2002; the estimated value was $487.5 million in 2004 or the equivalent of 11.4 per cent of the GDP of the OPT. Agriculture is the third largest employer in the OPT and in 2005 it employed about 117 thousand persons (15.2 per cent of the population) and double that if the informal work force of women and youth and marginalized sections of society are included. The contribution of the agriculture sector to the livelihood of the Palestinians is particularly important in the absence of viable economy (ARIJ 2007).

The majority of the Palestinian population, some 65 per cent, live in rural small villages which traditionally depend on agriculture. The majority of agricultural holdings in the OPT are small and sufficient to support one family. It is estimated that 88 per cent of the small holdings are held by one household. Consequently, the livelihoods of tens of thousands of Palestinian families are more than ever dependent on their ability to harvest their olives and market them locally, to

Israel and to the neighbouring Arab countries. Olive is one of the most important agriculture industries and is a significant source of income for the rural Palestinian population. The harm caused to this sector by the Israeli policies has harsh consequences for Palestinian livelihoods and standards of living.

The annual olive harvest varies from one year to the other; in 2000 126,000 tons of olive produced 27,000 tons of oil. In 2001 it was a weaker crop of 22,000 tons of olive for 5,400 tons of oil. It is estimated that about 92 per cent of the olive crop is intended for the production of oil and the rest for olive (B'Tselem 2008: 67). In 2007 the total value of agriculture produce in the OPT accounted for 25 per cent of GDP. The olive industry however, in the OPT supports some 70,560 households which accounts for 20 per cent of the total value of agriculture produce and accounted for 5.4 per cent of GDP for the same year (World Bank 2006).[3]

The importance of the agricultural and olive industry particularly has increased since the second Intifada in 2000 and the wide spread violent confrontations between Palestinians and the Israeli army. Over a hundred thousand Palestinian workers who had been employed inside Israel lost their jobs following the prohibition of entry for labourers into Israel. Consequently, this caused a significant reduction in income to many families as the work in Israel was the main source of earning. The GDP per capita in the Palestinian territories has declined by 36 per cent between 2000 and 2002. This highlighted the scale of the crisis and the need to start looking internally to improve the economic conditions of the population. As a result of the measures preventing Palestinian workers from entry into Israel, many Palestinians who had previously deserted their land to work as cheap labour in the Israeli market, had to return to agricultural work and cultivate the land. Many families started to lease land from large landowners. There has also been a return to old traditions where by families who do not own land harvest fruit and olives for a financial return and a share of the produce (ARIJ 2007).

Palestinian and Israeli human rights organizations have documented that the destruction of olive trees escalated with the outbreak of the second Intifada. The Palestinian Ministry of Agriculture reported that during the first year some 374,030 trees had been destroyed by the Israeli army and settlers. The impact of the loss of so many productive trees on the income of household and food security is significant especially considering that it takes up to ten years to have a full productive olive tree. As a result of such measures Palestinian farmers incurred losses of $300 million between September 2000 and June 2001 (Sarafa 2004 13).

Food security status has been negatively affected by the continued instability in the OPT and according to the World Food Programme, 49 per cent of the population in the OPT in 2006 had no food security. Olives, olive oil and bread are considered an important part of the daily diet especially for the low income families and, as such, the olive harvest is central to food security. This has become more critical as now thousands of Palestinian families in the OPT live under the

---

3   Interview conducted by the author with Adnan Ramadan, organizer of the international Solidarity Campaign with the Palestinians, 2010.

poverty line of less than two dollars per capita a day. It is estimated that in 2007 23.6 per cent of West Bank residents and 55.7 per cent of the Gaza Strip residents live below poverty line (UNDP 2011).

The olive harvest is considered by many poor families who do not own land as a vital means to support their income. Every season members of poor families, especially the young, will go and collect the remaining olives from the trees which the owners did not manage to collect. According to Sami Dauod,[4] a Palestinian farmer, 'it is a social custom to leave olives on the tree on purpose for people to come later and collect them which is seen as a way of supporting the poor'. It is also a custom that during the process of pressing olives ordinary passers-by can have a taste of the newly pressed olive oil.

It is commonly known that owners of olive orchards will support the poor in the community by supplying them with olives and oil. Some farmer will follow the rules of Zakat, one of the five pillars of Islam, which is a form of obligatory alms giving to be performed by Muslims which is usually 2.5 per cent of the total olive produce. However, others may use an arbitrary way based on good will, generosity and the success of that harvest season to determine the level of their giving.

In order to support the local olive economy, some local and international organizations have invested in micro finance projects to support women in establishing small businesses for pickling olives and selling them on to local markets. Local women's organizations and other development NGOs supported such programmes because they are recognized as successful projects and contribute to the local economy. The Palestinian Women's Technical Affairs Committees (PWTAC) argued that this is a way to fight poverty and to encourage women's independence and contribution to the household income (PWTAC 2009).

The confiscation of Palestinian land by the Israeli authority directly harms the income of many household families. Farmers face obstacles and restrictions in gaining access to enclosed land adjacent to the settlements. Often access is limited to a few times a year which, from a financial point of view, is not enough to make olive production profitable. One farmer described this challenge as follows: 'to pick 1,000 olive trees, for example, you need five or six adults working seven hours a day for 15–20 days.. we got a permit for one day, 16 October 2007, from 8:30 am to 2:00 pm. This was not enough time'. Stringent conditions set by the Israeli army limited the access for farmers to the enclosed land which means that they can't look after the land and trees which consequently affected the productivity (B'Tselem 2008: 67).

Israel has an obligation under international humanitarian law to compensate Palestinian farmers for harm caused by the occupying authority; however, Israel accepts very limited responsibility for compensation. The Israeli authorities have made the process of claiming compensation bureaucratic and have imposed many

---

4   Interview conducted by the author with Sami Dauod, director of the Palestinian Hydrology Group in Nablus, 7 November 2009.

conditions which resulted in the inability of farmers of enclosed land to receive compensation (B'Tselem 2002).

As a result of uprooting and damaging olive trees and other fruit trees many farmers lost their income from agriculture. The impact is greater for poor families and for families who own small orchards who make up the majority of farmers. In Jauyyous, a small Palestinian village in the West Bank, only 216 out 700 farmers have permits for some of the time, to work on their land behind the wall or adjacent to it. This resulted in half of the green houses in the village being abandoned and the inevitable reduction in productivity and income. This is in addition to the travel restrictions and transport of agricultural goods to the markets. A Palestinian activist organizing visits for international volunteers to support farmers during the harvest season argued 'the Israeli practices can be viewed as economic warfare and an assault on farmer's attachment to their land' (Ramadan 2010 Parry 2004).

Further challenges faced by the agriculture sector in Palestine in the last decade has resulted from an export driven agriculture development, including olives and oil, where the main focus is on export and meeting the demands of outside markets. Farmers in the Gaza Strip were encouraged by the Ministry of Agriculture at PNA and other international NGOs and given financial incentives to grow flowers and strawberries for export. In the West Bank international agencies such as the United States AID financially supported farmers to grow herbs and export oil for the European market. This policy pressured farmers to abandon traditional farming and move to new farming sectors which entailed a high risk approach. The fragile political situation in the OPT and the fact that Israel controls the border crossing and the movement within the OPT made the agriculture sector ever more vulnerable. The drive to move to export provided a stark example of such vulnerability when Israel decided to impose a siege on the Gaza Strip from 2006 and restricted import and export. This undermined the resilience of the population and increased the dependency on foreign aid and consequently exposed it to food insecurity and household coping mechanisms were unable to manage in the face of the occupation coupled with this embargo. The population was besieged and faced a dire threat to nutrition with little home grown resources and no trade permitted.

Conclusion

Despite all the difficulties Palestinian farmers' face, as this chapter highlighted, they continue to plant and harvest olive. Olive trees and harvest has political, economic, social and cultural importance to Palestinian society, it symbolizes connection to the land and manifestation of social values and traditions. In addition to the economic value and contribution to livelihoods it supported the resilience of the farmers to stay in their land and resist the occupation.

As described the majority of the Palestinians in the OPT live in rural areas and rely on household employment. There is a need for decision makers in the OPT to develop strategies to support livelihood programmes in the rural areas and especially the development of the agriculture sector as many local and international NGO's are already doing. Olive plantation has proved to be a source of income and human security in the region for hundreds of years. It appears to be imperative to increase the olive plantation and other sectors to create job opportunities and some sustainability in the rural areas. This could provide a significant improvement for the stability and viability of these people faced with the volatile political situation which also requires immense resilience in the face of the Israeli occupation.

This research has highlighted the need to support olive farming in adopting a more sustainable development driven approach rather than a trade and export driven strategy. Given that olive oil and olives are an essential part of the Palestinian diet such an approach will increase food security in the OPT. In the production process, there is a need to invest and improve olive pressing and raise the quality and reliability of locally produced food especially olive oil. The PNA allocated only 1.21 per cent of its budget in 2009 for agriculture and therefore investment in human resources and capacity building in research and training in olive farm management is critical to increase farmers' skills and awareness (Oxfam 2010).

List of Acronyms

OPT	Occupied Palestinian Territories
MOF	Ministry of Finance
SWC	Stop the Wall Campaign
NIS	New Israeli Shekel

References

Applied Research Institute-Jerusalem (ARIJ). 2007. *Status of the Environment in the Occupied Palestinian Territory*. Bethlehem: ARIJ.

Bornstein, A. 2008. Military occupation as carceral society: Prisons, checkpoints and walls in the Israeli Palestinian Struggle. *Social Analysis*, 52 (2)106–130.

B'Tselem, The Israeli Centre for Human Rights in the Occupied Territories. 2008. *Access Denied: Israeli measures to deny Palestinians access to land around settlements*. Jerusalem: B'Tselem.

B'Tselem, The Israeli Centre for Human Rights in the Occupied Territories. 2002. *The Performance of Law Authorities in Responding to Settler Attacks on Olive Harvesters*. Jerusalem: B'Tselem.

Cheri, B. 2001. *Washington Report on the Middle East Affairs*, 20 May/June, 94–110.

Darweish, M. 2010. Human rights and the imbalance of power: The Palestinian-Israeli conflict, in *Human Rights and Conflict Transformation: The Challenge for Just Peace*, edited by V. Dudouet and B. Schmelzle. Germany: Berghof Handbook Dialogue Serious 9, 85–93.

The Ecumenical Accompaniment Programme in Palestine and Israel (EAPPI). 2010. *An unjust settlement: A tale of illegal Israeli settlement in the West Bank.* EAPPI.

Hass, A. 2006. It is not the Olive trees, *Haaretz*, 11 January.

The Palestinian Human Rights Monitor. 2001. *Criminal negligence? Settler violence and state inaction during the Al-Aqsa Intifada*, 5(2).

Khalaf, A. 2009. Olive and fighting poverty, *Woman Voice*, Palestinian Women Technical Affairs Committees (PWTAC), 22 October, 4. Arabic

Lynfield, B. 2000. Another casualty of war: trees. *The Christian Science Monitor*. 12 August.

Parry W, 2004. Washington Report, October.

Oxfam, 2010. The Road to Olive Farming: Challenges to Developing the Economy of Olive Oil in the West Bank. Oxfam [Online]. Available at: http://www.oxfam.org/sites/www.oxfam.org/files/the-road-to-olive-farming_0.pdf [accessed: 15 January 2011].

Sarafa, R. 2004. Roots of conflict: Felling Palestine's olive trees. *Harvard International Review*, Spring, 13–15.

Shehadeh, R. 2008. *Palestinian Walks: Notes on a Vanishing Landscape*. London: Profile books.

Stop The Wall Campaign 2010, Interview with Director, Ramallah.

World Bank. 2006. Brief Overview of the Olive and the Olive Oil Sector in the Palestinian Territories [Online]. Available at: http://web.worldbank.org/WBSITE/EXTERNAL/COUNTRIES/MENAEXT/WESTBANKGAZAEXTN/0,contentMDK:21693806~pagePK:141137~piPK:141127~theSitePK:294365,00.html [accessed: 15 January 2011].

UNDP. 2011. *Human Development Report 2009/10, Occupied Palestinian Territories: Investing in Human Security for Future State*, Jerusalem: UNDP.

Zertal, I. and Eldar, A. 2007. *Lords of the Land: The War Over Israeli's Settlements in the Occupied Territories, 1967–2007*. New York: Nation books.

Chapter 14

Practical Action in North Darfur

Liam Morgan and Barnaby Peacocke

Introduction

The conflict in Darfur began in 2003 as an uprising among disaffected tribes. As the rebellion grew the government launched a lethal counter-insurgency, arming and supporting tribal militia to break the rebellion and its support base in the rural heartlands. The subsequent violence and destruction resulted in the deaths of tens of thousands of people, and displaced two million (Human Rights Watch 2011). The root causes of the conflict are multifaceted, and far more complex than the simple dichotomy of 'Africans' versus 'Arabs' often portrayed through the world's media (Young et al. 2009). It has continued with fluctuating levels of intensity since, with devastating impacts on rural livelihoods and agricultural productivity.

North Darfur is the most arid of the three Darfur states. It is the one most prone to drought, sporadic rainfall and famine. Food security has been especially hard hit by the conflict. While food shortfalls have to some extent been filled by food aid, which has helped keep people alive, stabilize prices and keep markets functioning (Jaspars and Maxwell 2009),[1] it has also created dependencies and disempowered local people, especially in their ability to sell staple crops through local markets (UNDP 2010). Yet, despite the hardships faced, most Northern Darfuris have managed to stay in their rural communities. Some have even been able to increase their levels of agricultural production and respond to growing market demand from expanding urban centres and displacement camps.

This chapter outlines the role and experiences of the international non-governmental organization (INGO) Practical Action in supporting agricultural production among rural communities in North Darfur over the conflict period. It will shed light on the organizational and programmatic response to the shifting context and provide insights into some of the key success factors behind the programme approach, as well as the challenges faced and lessons learned.

---

1   Although, such is the perverse nature of world food markets, in 2008 Sudan exported 300,000 tons of sorghum to Saudi Arabia for animal feed, roughly the same figure that was imported for food aid (Dorosh and Subran 2009).

Practical Action's Pre-conflict Approach to Food Production

The Interdependent Agricultural Package

Practical Action has been working in North Darfur since 1987. It began by providing technical assistance on food production to Oxfam GB following the 1984–5 famine. In 1994 it established its own office in El Fasher, the state capital. At this time the focus was on working with local farmers to develop a range of agricultural techniques to reduce drought risks and address their accessibility to poor people. Much of the technical emphasis behind this work explored a shift in cropping practices away from the degraded and unproductive *goz*, and into the high-potential wadis.[2] A donkey plough was designed and tested capable of tilling the heavier wadi soils. Vegetable production was expanded and early-maturing open-pollinated millet and sorghum varieties introduced from other parts of Darfur and Sudan. Rainwater harvesting techniques such as crescent terraces and earth dams adapted from existing local technologies were established in the wadis. Dams were modified with sluice gates to flush out silt deposits and reduce the risks of collapse during peak flows. These technologies were backed up by generic inputs such as seeds, locally made hand tools and the building of physical assets such as community grain, seed and tool banks. To support these provisions, a cadre of village extension agents (VEAs) was trained to disseminate agricultural knowledge and skills within their communities, promote farmer experimentation and build links between isolated communities and government technical services.

Supporting the Establishment of Representative Rural Civil Society

For there to be real sustainability and opportunities for wider take-up it was important to support technological innovation by strengthening the organizational and management capacities of rural communities. Practical Action adopted Oxfam GB's approach, which encouraged the formation of community-based organizations (CBOs) including Village Development Committees (VDCs), Women's Development Associations (WDAs) and Blacksmith Associations. Each CBO would represent a cluster of villages supporting an average of 200 households. CBO formation typically included an executive and management committee. Members received training and support for the prioritization, planning and implementation of local activities. By the end of 2002 Practical Action had helped establish 28 of these CBOs providing outreach to over 50,000 households.

---

2   *Goz* lands are ancient stabilized sand dunes. The soils are light and easy to cultivate by hand, but do not retain water and are only suitable for short-season millets. Wadis are wide clay plains that run alongside a seasonal river bed that may flow two to five times a year. A wadi may range from 0.5 to 3 kilometres wide. The soils are fertile clays, water-retentive and highly productive but difficult to cultivate.

Practical Action's Experience in Working through Protracted Conflict

Direct Early Impacts on Agricultural Operations and Activities

Shortly before the start of the conflict in 2003 Practical Action had begun to implement a food security project in partnership with 30 CBOs under European Union funding. Deteriorating security forced a withdrawal from 12 severely affected CBO areas that had suffered burnings, displacement and in one example the appropriation of a recently constructed grain bank by a rebel faction in need of a headquarters. Other grain banks had become targets for looting, prompting Practical Action to recommend villagers return to storing their grain underground or in their homes.

Increased travel restrictions led Practical Action to focus continuing activities across rural village clusters closer to El Fasher town. Target groups included both resident and displaced communities hosted by the villages that had been left outside mainstream humanitarian efforts. Even here the agency struggled to visit or get supplies to villages due to insecurity and travel restrictions in areas controlled by the Sudan Liberation Army (SLA). This situation quickly led CBO members to begin travelling to El Fasher to receive training, advice and inputs to take back to their villages. But this strategy was far from ideal. Nascent war economies such as military checkpoints induced higher travel costs and this informal economy began to proliferate as rebel groups splintered.

Operational Response to the Changing Context

As the Darfur conflict ebbed and flowed and rural access became nearly impossible, Practical Action staff in El Fasher recognized it needed to change the way it operated. The seeds of a solution were first sown in 2003 when a group of village-level WDAs came together to form the Women's Development Association Network (WDAN). Although its operational scope was initially limited to helping women in urban centres, this realignment provided an impetus for other CBOs to form new umbrella networks.

In 2005, after two years of intense violence in rural areas, a group of 18 VDCs visited the Practical Action office in El Fasher to discuss new ways to coordinate activities, and voice concerns that Practical Action might close its El Fasher office due to the insecurity.[3] In a spontaneous act, the idea was raised by one VDC representative that if Practical Action was unable to sustain its outreach they should themselves form an organization capable of linking their outlying villages to support and resources in El Fasher town. The result was the establishment of the El Fasher Rural Development Network (RDN). With Practical Action refocusing its operational approach to support the network, the RDN quickly developed its

---

3   Interview with RDN senior staff member, North Darfur, July 2011.

operational capacity to reach rural areas through its VDC membership and network of village extension agents and paravets.

Formation of the RDN was followed in 2006 by a second group of VDCs who formed the Market Network (since renamed the Voluntary Network for Rural Help and Development, VNRHD). Together with WDAN, the three networks began by taking on tasks such as paying and monitoring the work of contractors. This emergent strategy explicitly acknowledged that state, UN and INGO actors lacked the capacity or will to provide services to rural communities in the face of recurrent insecurity. It also provided a means of empowering civil society. Once established, the networks began to extend their membership and secure their own funding from UN bodies for small-scale projects, with Practical Action overseeing the administration of funds on their behalf.

Concurrently, Practical Action supported the formation of a Blacksmith Association in El Fasher town. This brought together blacksmiths who had fled rural areas during the conflict with those already based around the market. Working as an association enabled the blacksmiths to organize themselves to negotiate major contracts producing agricultural tools and donkey-drawn ploughs for major UN and INGO humanitarian agencies while gaining economies of scale in both inputs and production (Abdelnour 2010).

Programmatic Response to the Changing Context

Throughout the crisis Practical Action continued to promote a number of pre-conflict agriculture-based interventions, including an expansion of earth dams. Even during the peak of the 2003 counter-insurgency Practical Action successfully worked with the communities around Umm Bronga south-east of El Fasher to create an entirely new dam. In 2004 the dam flooded 2,000 hectares, providing food to around 45,000 people, and year-round work for more than 10,000 more, mostly those displaced by war. While the original approach to dam construction was to use localities like Umm Bronga as models to encourage wider investment and support by development agencies, government departments, banks and donors, this wider take-up has only recently secured greater support as the security situation has become more predictable. As a result, in the absence of wider uptake, dam construction continued as a core part of programme activities throughout the crisis period.

Counter-intuitively, windows also began to open for a shift towards long-term work relevant to the underlying conflict. The success of the networks in accessing funds meant they increasingly took on the agricultural support work themselves. This in turn made space for Practical Action to begin to target wider issues of natural resource management including the promotion of sustainable rangeland and forest use practices. Bringing together the growing capacities of civil society groups and technical requirements of sustainable resource use allowed Practical Action to begin to address the governance structures and processes that had traditionally controlled natural resource access and use by farmer and pastoralist groups. From the 1970s these institutions had been sidelined and traditional approaches to conflict resolution politicized (see, for instance, Young et al. 2009). And while initial attempts to introduce peacebuilding elements were blocked in 2007, the language considered provocative, by 2009 the way was open to introduce consensus-building approaches for decision-making and to support traditional dispute resolution methods over the provision of services and allocation of resource access rights.

Practical Action's approach to consensus building is known as Participatory Action Plan Development (PAPD). Tested and developed in Bangladesh for access to common property resources (Lewins, Coupe and Murray 2007), PAPD incorporates a 'structured and repeatable set of activities that helps local people identify key problems and constraints together with realistic opportunities to address them' (Taha et al. 2010: 2). A key part of the approach is to facilitate dialogue between a range of primary stakeholder groups over resource access rights and controls. Tested over the period 2009–11, PAPD has been used in Darfur to resolve a violent dispute along ethnic lines over control of a village WDA, and a conflict between two villages over the locating of a traditional water reservoir (Arabic *haffir*).

Reflections on Practical Action's Experience of Promoting Agriculture during Conflict

The Evolving Role of Community-based Organizations in Strengthening Food Security

The formation of three civil society networks was initially seen by Practical Action and the network members as a means of providing an institutional framework for bridging the gap between El Fasher and outlying rural areas cut off by insecurity and access restrictions imposed by government and rebel troops. In the earlier days of their formation the networks provided a form of remote management. Today they are autonomous organizations that do not see their role as enabling INGO, UN or government organizations to increase their operational efficiency unless it suits their own purposes. That said, while the formation of networks is the outcome of social innovation from within the communities themselves supported by Practical Action's desire to empower civil society in Darfur, it is highly unlikely they would have developed a similar, rapid, change trajectory or that Practical Action would have been so supportive if it had not been for the rapidly changing context precipitated by the conflict.

The three CBO networks supported by Practical Action have made significant contributions to increasing the food security of their membership in rural areas. During the early years of the conflict, fear of attack and roadblocks severely restricted farming, even in villages that were not directly affected (Jaspars 2010). But because the CBO network membership was located in both government- and rebel-controlled areas, the networks had a comparative advantage over INGOs and UN agencies in negotiating secure movement for farm and market access (Jaspars

2010). The networks were also helped by the fact that their membership is drawn from across Darfur's multi-ethnic base. Network representation has managed to balance the need to maintain contact with government and opposition factions, and the membership bridges different political groups. But they have managed to stay out of the conflict itself. As a result the networks have on the whole managed to provide a set of spatial, social and political bridging functions allowing them to engage in local mediation and conflict resolution activities (Jaspars 2010). As network support led to improved agricultural productivity, market integration became increasingly important and Practical Action has supported the networks in the collection of marketing and price information to serve their members rather than the monitoring needs of humanitarian agency and government decision-makers (Buchanan-Smith and Fadul 2008).

While most agencies concentrated their work in the IDP camps, the networks began to take advantage of donor attempts to increase fund allocations to rural areas. These opportunities increased when 13 INGOs were expelled by the Sudanese government in 2009. By the summer of 2011 the three CBO networks had accessed nearly US$4 million from a range of UN and NGO donors (Practical Action 2011). As a result, all village members of the RDN receive more development inputs now than before the conflict (Jaspars 2010) and the UNDP have declared that INGO-formed networks have demonstrated 'remarkable resilience' during the conflict period (UNDP 2009).

This funding phenomenon also occurred in El Fasher town, where the Blacksmith Association received several contracts upwards of US$250,000 from organizations like the UN's Food and Agriculture Organization (FAO) and the International Committee of the Red Cross (ICRC) for the provision of locally adapted agricultural tools for humanitarian programmes. This success has realized some significant social benefits. Blacksmiths were originally from a highly marginalized caste, and they have become increasingly accepted by wider society due to their economic rise, and growing public recognition of their role as 'key drivers of agricultural improvement in Darfur' (Abdelnour 2010: 23).

Yet despite these successes, there is a need to be cautious. The nature of conflict can have perverse impacts that quickly reverse any successes. For instance, while WDAN members successfully promoted a range of agricultural activities in the area of Kafout during 2006, these were rapidly reversed by a new wave of displacement in 2007.[4] In 2009 there was a violent split along ethnic lines which led to the break-up of one village WDA group with no previous history of ethnic tension. And while this clash has since been resolved, there are examples of disputes within at least five village clusters over inequitable distribution of resources between villages overseen by the VDCs. Furthermore, none of the three

---

4   Women make a considerable contribution to agriculture in Darfur, and are engaged in an estimated 75 per cent of sowing, harvesting and weeding activities (Fidiel 2001, cited in Abdelnour 2010).

civil society networks has yet held an election for the management or executive since its inception, which may carry risks of future challenges over self-interest.

Nevertheless, despite difficulties the networks have continued to evolve, expand and serve increasing numbers of highly supportive members. Networks have effectively increased their social capital by consolidating and improving relations between communities and livelihood groups, and their political capital by engaging and influencing state and non-state actors and market institutions. The three networks now include over 150 CBO members, representing at least 170,000 individuals. This growth has been fostered by the networks expanding their CBO base organically. Although there are obvious benefits to support this expansion, it is important that the networks do not grow faster than their operational capacity permits. For this reason Practical Action has lobbied UN and INGO agencies in North Darfur to work with existing networks established by other agencies during the 1990s in Kutum, Kabkabiya, Dar El Salam and Umm Kedada that lie beyond its current operational reach.

Engaging with the State to Promote Agriculture during Conflict

Village extension agents (VEAs) have provided a vital service to their VDCs throughout the conflict (see Coupe and Pasteur 2009). The VEAs' concentration on knowledge services, their election by community members and oversight by the VDC executive can in some ways be seen as a conflict-sensitive practice when compared with the provision of material goods (Jaspars and Maxwell 2009). VEA services are provided free to community members in return for a donkey and cart on loan from the VDC. The VEAs are trained by Ministry of Agriculture (MoA) staff through a one-month programme in El Fasher town. Like the networks, they have built on these linkages to provide a bridging function between outlying villages and government technical staff (Coupe and Pasteur 2009).

These linkages with MoA technical staff are important both operationally and politically. They provide technical back-up that Practical Action cannot, and have supported the programme objectives from its inception and through the crisis. Practical Action has supported the MoA in conducting several pre-harvest and post-harvest studies providing up-to-date information on food deficits and evolving priorities, and engaged in joint technical broadcasts on El Fasher radio on a range of agricultural topics. In 2010 Practical Action worked with the MoA in setting up a seed fair in El Fasher town. This was the first fair since the start of the conflict and was aimed at promoting varietal diversification and testing by farmers. Practical Action's low-level relationship with the state, focused on relatively unpoliticized technical and long-term development engagements, enabled the agency to support communities in food and agriculture, water, forestry and markets and avoid the expulsion of 13 mainly humanitarian agencies in early 2009.

Increasing Agricultural Productivity through the Provision of Earth Dams

It is difficult to measure just how effective Practical Action's interdependent agriculture package has been due to the presence of wider humanitarian interventions. But some examples of productivity increases have been so significant that their cause is self-evident. The provision of earth dams in wadis has allowed surrounding villages to hugely increase their productivity, well beyond pre-conflict levels. In one account, independent evaluators from the European Union estimated in 2006 that as a result of the shallow flooding of 5,000 hectares of land across three dams, 'food production per household had increased by between 500 and 800 per cent' (EU 2006).

While some agencies have argued that the security situation in Darfur and other conflict situations means that substantial construction projects such as earth dams should not be readily undertaken (for example, Tearfund 2007), Practical Action's experience has been to show these types of projects are feasible. Nine dams have been constructed since 2003 including seven new structures, one rehabilitated dam and one completely rebuilt. Despite the crisis, none were interrupted. This owed to a common recognition that dams directly address food production needs and involve strong local ownership, enabling communities to negotiate support from the different armed factions. This sense of ownership was particularly strong in localities like Umm Bronga where communities had closely participated in the dam construction through household monetary contributions, and the supply of labour and masons.

Fidiel (2005) argues that the impacts of production increases around dams have on the whole been equitable because land allocations have followed established norms overseen by village headmen (sheikhs). Villages with dams have also provided a strong 'pull factor' attracting internally displaced people to leave camps and settle as tenants near the wadi, or to work as migrant labourers, thereby providing essential employment opportunities where few were available (Practical Action 2009). This influx of migrants also included some professionals, resulting in the reopening of village schools in some localities.

It can also be argued that the combination of ongoing conflict and the use of traditional institutions to oversee access arrangements have helped Practical Action avoid some of the risks of external elite capture that often dominated infrastructure investment programmes in Sudan (see, for example, Verhoeven 2011). For example, during the 1980s in the neighbouring state of Kordofan, government-backed Khartoum elites were allowed to buy up wadi lands and take over ownership of new structures while excluding local communities (Haaland 1991).

Yet, despite some obvious successes, one dam in particular has been at the centre of tensions between neighbouring communities. Downstream communities believe that the dam at Abu Digeis village prevented the wadi flow from flooding their fields during the 2009 rains. They disagreed with the Water Services Board that wadi flows had been limited that year by drought. Under the supervision of a local leader they destroyed part of the dam wall. Jaspars (2010) claimed the problem stemmed from inadequate consultation between affected groups. Others argue that the problem originated from differences between neighbouring *omdas*, traditional leaders overseeing the different village clusters involved in the dispute. Either way, the example highlights the critical need for intense negotiations between upstream and downstream communities. This is also the case where dams are located on or near to traditional pastoralist routes (Arabic *massar*). In the former case they can impede the route, while in the latter they can quickly provide a draw to migrating herds with associated opportunities, for example feeding and watering arrangements. Both can potentially lead to increased tensions between farmers and pastoralists.

Farming, Nomadic Pastoralism and the Challenge of Impartiality

Leading on from the previous section, it is important that one takes a step back when looking at the promotion of agriculture in North Darfur. Agriculture and pastoralism, or a combination of the two, are the major livelihood strategies followed in rural areas. These livelihood systems have traditionally benefited from a symbiotic relationship that has broken down in recent years and fed the conflict dynamics. The Northern Rezeigat, an Arab tribe who constitute part of the pastoralist *abbala* (nomadic camel-herding groups), is historically one of the main tribes in Darfur not to have been awarded its own homeland or *dar*. Since independence, government and international support to the region has on the whole neglected nomadic groups (De Waal 2005, Ghaffer and Ahmed 1987), a situation that has generally continued since the onset of the conflict (Young et al. 2009).[5] So the Northern Rezeigat and similar nomadic pastoralist groups can be seen as a marginalized element within a marginalized region. In the absence of any representation among the rebel groups at the outbreak of the conflict in early 2001, the Khartoum government was able to rapidly recruit and arm members of the Northern Rezeigat into the *janjaweed* militia to join the counter-insurgency.

Such social, livelihood and political tensions between pastoralists and sedentary populations are by no means resolved. In a recent sign of continuing pressure on pastoralist livelihoods, two hundred *abbala* herders entered El Fasher locality in December 2010 three months earlier than agreed under customary agreements.[6] The insecurity situation at that time also caused them to travel in armed groups, fostering fear among local farming communities. Ensuing clashes left at least three dead.[7]

Many of the troubles between the groups can be traced back to the droughts of the 1980s. These forced many from the Zaghawa tribe to migrate from their *dar* in north-west Darfur and establish villages south of El Fasher. The growth of these

5   The *abbala* population in North Darfur is estimated at 350,000 out of a total of approximately 1.5 million.

6   There is a time around March when farmers have finished harvesting their crops that pastoralists are permitted to enter and feed on crop residues – this is known as '*Talaig*'.

7   Interview with NGO staff member, North Darfur, May 2011.

villages and expansion of their agricultural lands led to the complete blockage of one of the main north–south migratory routes. Other *massarat* along the Jebel Mara have been encroached by farmers enclosing their land as they sought to increase the commercial production of vegetables and tobacco. Wider tensions have been exacerbated by farmers choosing to sell crop residues that had previously been grazed by the nomads under customary access arrangements (Young et al. 2005, Pantuliano 2007). And the ad hoc growth of market-orientated agriculture along some of the main wadis using diesel pumps and shallow wells has impeded several pastoralist migration routes (Mohamed 2004).

In response to these pressures Practical Action has increasingly explored ways of engaging pastoralist groups in programme implementation. In 2006 a number of nomads were trained as paravets capable of vaccinating animals, and one of the main reasons behind promoting the use of wadi lands was to free up *goz* areas for animal pastures. In some localities this has happened. Up to 50 per cent of *goz*-soil lands have been abandoned for wadi cultivation.[8] But in other localities farmers have continued to cultivate the *goz* while also expanding production into the wadis. In these areas and in those without any wadi the emphasis has been to promote the use of early-maturing varieties that help farmers harvest their *goz* crops before animal herders arrive during the seasonal migration and to support negotiations to allow grazing on crop residues.

Practical Action has also tried to draw pastoralist groups into programme planning. In early 2011 a 'Greening Darfur' conference was held in El Fasher to establish a public platform for farmer–pastoralist debate that had been lacking. The focus was on identifying common development opportunities and constraints. Building on this dialogue, Practical Action's work began to explore ways of incorporating nomadic pastoralist groups into local decision-making through joint development planning. This process used the PAPD consensus-building framework already tested in Darfur to bring farmers and pastoralists together to air their viewpoints in a sequential planning process. One result of the conference discussions was for Practical Action to start training pastoralist representatives as PAPD facilitators with the aim that they work alongside a similar cadre that had by that time been established within the three networks. In addition, a small number of pastoralists have begun to participate in village-level consensus-building sessions taking place in communities located near migratory routes where there are known tensions over resource access and control. In turn, these local-level sessions have started to underpin a constructive dialogue at the North Darfur state level over the reconfirmation of *massarat* migratory routes and associated pastures designated for pastoralist use (Arabic *sineyya*). With increasingly explicit donor and government support to these processes, it is hoped they will provide pastoralists with a practical means of addressing their sense of historical exclusion.

---

8   Interview with VNRHD senior staff member, North Darfur, August 2011.

The Way Forward

Agriculture has provided the bedrock of Practical Action's work in North Darfur since before the conflict. A clear focus on agricultural productivity has legitimized the agency's presence across political factions and enabled the programme to support the food security of tens of thousands of people. This has been supported by work addressing the social and organizational capacities of rural communities and their representation in El Fasher among government agencies, humanitarian and development organizations, and donors. It is unlikely that a focus on either social or technological processes would have worked alone. Village-level groups formed to support the take-up and expansion of agricultural production methods underwent a social innovation to form civil society networks that have since driven the rapid expansion of community networks and blacksmiths' groups. The pressure to innovate and expand membership and influence at local and state level would probably not have been as strong in the absence of conflict. It is to be seen whether these pressures may also drive efforts to foster intercommunity reconciliation and stability.

The foundation of most of these changes was Practical Action's prior engagement in the region and its tactical decision to sustain a long-term approach as the conflict began to spill over during the period from 2003 to 2005. One of the key reasons behind this decision was that the programme was led by a local Sudanese team headed by Darfuris. In effect these individuals provided a natural extension of civil society influence into Practical Action itself. But it is also because of the sustained support of a small number of donors including EC Food Security, Christian Aid and UNDP's Darfur Community Peace and Stability Fund that the long-term approach could be maintained.

Despite some significant successes, there are many challenges ahead. The difficulties of responding to complex interactions between livelihood strategies overlayered by ethnic differences in Darfur should not be underestimated. Those who understand these complexities and who are best placed to navigate them are the rural communities themselves. In many ways the three CBO networks have shown themselves willing and capable of addressing some of the most pressing livelihood and conflict issues. They continue to demand financial and technical support and will need political backing if they are to help tackle some of the larger questions facing the state over land and water access. Over time, current trends suggest the networks will increasingly oversee joint development plans at the village level. However, efforts to build consensus between farmers and pastoralists over the location and use of new dam structures, animal migration routes and designated pastures are complex and affect a number of neighbouring territories, ethnicities and interests. As a result they may continue to rely on independent facilitation and support. Nevertheless, the clear voice from the Greening Darfur conference was that civil society groups do want to build consensus over resource-use arrangements and improvements, and this can in turn provide practical opportunities for building stability from the bottom up.

References

Abdelnour, S. 2010. *Forging Through Adversity: The Blacksmiths of North Darfur and Practical Action.* Growing Inclusive Markets Case Study No. B057. New York: UNDP.

Baumann, P. 2000. *Sustainable Livelihoods and Political Capital: Arguments and Evidence from Decentralisation and Natural Resource Management in India.* London: ODI.

Buchanan-Smith, M. and Fadul, A.A. 2008. *Adaptation and Devastation: The Impact of the Conflict on Trade and Markets in Darfur.* Medford, MA: Feinstein International Center, Tufts University [Online]. Available at: http://fic.tufts.edu/?pid=83.

Buchanan-Smith, M. and Jaspars, S. 2007. Conflict, camps and coercion: the ongoing livelihoods crisis in Darfur. *Disasters*, 31(S1), S57–S76.

Coupe, S. and Pasteur, K. 2009. *Food and Livelihood Security.* Lesson Learning Study Including Community Extension Sudan Country Study Report. Rugby: Practical Action.

De Waal, A. 2005. *Famine That Kills.* Revised edition. Oxford: Oxford University Press.

Dorosh, P. and Subran, L. 2009. *Food Aid, External Trade and Domestic Markets: Implications for Food Security in Darfur.* Contributed paper prepared for presentation at the International Association of Agricultural Economists Conference, Beijing, China, 16–22 August 2009.

EU. 2006. Monitoring report Sudan. Re-establishing food self-reliance among drought affected populations of Northern Darfur, Sudan. MR-01655.01-06/04/06 Project Number: FOOD/2001/048-427.

Fidiel, M.M. 2001. *The Case of Developing the Donkey Drawn Plough in North Darfur, Western Sudan.* Khartoum: ITDG Sudan.

Fidiel, M. 2005. *Building Small Scale Water Harvesting Dams: The Experience of Intermediate Technology Development Group, North Darfur State – Western Sudan.* ITDG [Online]. Available at: http://practicalaction.org/sudan/docs/region_sudan/water-harvesting.pdf.

Ghaffer, A. and Ahmed, M. 1987. National ambivalence and external hegemony: the negligence of pastoral nomads in the Sudan, in *Agrarian Change in the Central Rainlands: Sudan, A Socio-Economic Analysis*, edited by M. Salih. Uppsala: Nordiska Afrikainstitutet, 1987.

Haaland, G. 1991. Systems of agriculture in western Sudan, in *The Agriculture of the Sudan*, edited by G.M. Craig. Oxford: Oxford University Press, 230–51.

Hamid, A., Salih, A., Bradley, S., Couteaudier, T. El Haj, M., Hussein, M. and Steffen, P. 2005. *Markets, Livelihoods and Food Aid in Darfur: A Rapid Assessment and Programming Recommendations.* UN/FAO/USAID.

Human Rights Watch (HRW). 2011. *Darfur in the Shadows.* New York: Human Rights Watch.

Jaspars, S. 2010. *Coping and Change in Protracted Conflict: The Role of Community Groups and Local Institutions in Addressing Food Insecurity and Threats to Livelihoods*. London: Overseas Development Institute.

Jaspars, S. and Maxwell, D. 2009. *Food Security and Livelihoods Programming in Conflict: A Review, Network Paper*, 65. London: Humanitarian Practice Network.

Jaspars, S. and O'Callaghan, S. 2008. *Challenging Choices: Protection and Livelihoods in Darfur*. London: Overseas Development Institute.

Lewins, R., Coupe, S. and Murray, F. 2007. *Voices from the Margins: Consensus Building and Planning with the Poor in Bangladesh*. Rugby: Practical Action Publishing.

Mohamed, Y. 2004. Land tenure, land use and conflicts in Darfur, in *Environmental Degradation as a Cause of Conflict in Darfur*. Conference Proceedings, University for Peace, Khartoum, December 2004.

Pantuliano, S. 2007. *The Land Question: Sudan's Peace Nemesis*. London: Overseas Development Institute.

Practical Action. 2008. *Greening Darfur Project Document*. Khartoum: Practical Action.

Practical Action. 2009. *Practical Action Sudan – Annual Highlights 2008–2009*. Khartoum: Practical Action [Online]. Available at: http://practicalaction.org/sudan/docs/region_sudan/.

Practical Action. 2011. *Greening Darfur Conference Report*. Khartoum: Practical Action.

Taha, A., Lewins, R., Coupe, S. and Peacocke, B. 2010. *Consensus Building with Participatory Action Plan Development in Sudan: A Facilitator's Guide*. Khartoum: Practical Action and Christian Aid.

Tearfund. 2007. *Darfur: Relief in a Vulnerable Environment*. Tearfund [Online]. Available at: www.Tearfund.org/darfurenvironment.

UNDP. 2009. *Darfur Livelihoods Programme: Mapping and Capacity Assessment of Civil Society Organizations (CSOS) in Darfur*. UNDP.

UNDP. 2010. *Enhancing Livelihood Opportunities and Building Social Capital for New Livelihoods Strategies in Darfur*. UNDP Sudan [Online]. Available at: http://www.sd.undp.org/projects/cp8.htm [accessed: 30 June 2010].

Verhoeven, H. 2011. Climate change, conflict and development in Sudan: global neo-Malthusian narratives and local power struggles. *Development and Change*, 42(3), 679–707.

Young, H. 2007. *Livelihoods, Migration and Remittance Flows in Times of Crisis and Conflict: Case Studies for Darfur, Sudan*. London: Overseas Development Institute.

Young, H., Osman, A., Abusin, A., Asher, M. and Egem, O. 2009. *Livelihoods, Power and Choice: The Vulnerability of the Northern Rizaygat, Darfur, Sudan*. Medford: Feinstein International Famine Center, Tufts University.

Young, H., Osman, A., Aklilu, Y., Dale, R., Badri, B. and Fuddle, A. 2005. *Darfur – Livelihoods Under Siege*. Medford: Feinstein International Famine Center, Tufts University.

Glossary

abbala		Camel-herding pastoralists

dar		Tribal lands allocated by the Darfur Sultanate and later reconfirmed by the British Protectorate under the Native Administration

goz		Stabilized sand dunes easily cultivated by hand and used for millet production

haffir		Medium-sized water reservoir of up to 25,000 m³ capacity

janjaweed		Armed militia in Sudan

massar		Traditional route used by pastoralists to move between seasonal grazing lands (pl. *massarat*)

omda		Traditional leader overseeing village sheikhs within a dar

sineyya		A circular pasture area covering approximately 5 square kilometres at points along each migratory route designated for use by pastoralists

wadi		Clay depressions with a central watercourse that may flow up to five times a year

Chapter 15

Agricultural Information amid Conflict: For Whom and for What?

Ian Christoplos

Why Information Is Important for Farmers and for Those Supporting Recovery Processes

This chapter describes how in Afghanistan, Bosnia-Herzegovina and the former Soviet republics of Central Asia, relevant aspects of institutional development for agricultural information have been supported, skewed or ignored in recovery efforts. Particular attention is paid to analysing how misconceptions arise about what farmers want, who farmers are, and where there are market-related opportunities for recovery. The chapter provides direction for applying current understanding of agricultural advisory service provision in promoting more realistic, sustainable and equitable agricultural recovery efforts.

Agricultural recovery after conflict must be understood in the context of how farmers and other agricultural actors struggle to understand and respond to the changes that have occurred (and which are still under way) in markets, technologies, environmental conditions and the shifting roles of the state, farmer organizations and private sector. It is also about recognizing how the population that farmed before the conflict may no longer be willing or able to return to their past livelihoods. Institutional sources of information have usually been altered and often weakened in the course of the conflict. Asymmetrical access to information and other agricultural services may even have aggravated the conflict itself. Both farmers and those providing them with services undergo profound change in the course of a conflict.

Rural people need access to information if they are to rebuild, or at least stabilize, their livelihoods after conflict. They need to understand changes under way in the markets that have been transformed by war or by global shifts during the course of the war. After years of relative isolation, they may need to learn about new technologies and crop varieties and how these are impacting on their competitiveness and ability to manage climate uncertainty. They need to understand changes in consumer preferences, food standards and retail systems if they are to access markets beyond their villages (and in some cases, even within their villages). They need to understand how global food price instability is likely to affect how much they will receive for their crops and how much they will have to pay for food when their own production runs out. They need information to consider how changes in land, water and natural resource tenure systems have emerged either due to the conflict, or when new commercial actors from distant countries make investments that radically change their local agrifood systems. They need to reconsider what information services they are likely to receive from their national government, local government, aid agencies, farmer organizations and the private sector. Above all else, many rural people in post-conflict societies are trying to gather information so as to make informed decisions about whether to try to rebuild their livelihoods around agriculture, or if it is better to pack up and leave for a new life in the city.

It almost goes without saying that the vast majority of this information is rarely available, and what is available is highly fragmented. Rural people must look for clues that can help them to make sense of their situation. They gather information from neighbours, the media, remnants of the governmental agricultural extension service, NGO staff, large farmers with whom they are engaged in patron–client relations, traders, shop owners, and investors making offers to buy land or contract farmers to produce for them. Warlords, their government or investors may be ordering them what to grow if they are to retain control of their land.

This messy situation is not unique to farmers in post-conflict contexts. Similar challenges exist for the rural poor in relatively stable contexts. But in a post-conflict context these factors tend to come together in a more pressing manner. Their commercial systems have been grossly distorted due to breakdowns in systems by which they used to access export markets, growth of new markets due to the war (such as illicit crops), aid programmes that have heavily subsidized certain inputs (such as seeds) or marketed the farmers' products, or due to the temporary (perhaps permanent) collapse of government services. False assumptions about the future intentions of rural populations and their prospects for maintaining farming-based livelihoods frequently lead to misplaced priorities in post-conflict support for these services. It is uncertain whether many of today's rural poor are likely to be farmers in the future, given prevailing trends toward larger farms, increasing landlessness and unattainable market standards for agricultural products. Post-conflict NGO support is frequently predicated on expectations that the advisory services they provide can be 'handed over' to the government, farmer organizations or the private sector, none of which may have intentions to support smallholder farming. This chapter draws conclusions about the importance of recognizing and responding to these trends and institutional realities, rather than searching for silver-bullet solutions.

The importance of information in recovery efforts is related to the growing emphasis on supporting the capacities of conflict-affected people to pursue and preserve their livelihoods. It is now recognized that a temporary injection of food aid will not result in sustained reduction of hunger. Aid modalities – including humanitarian assistance – are being reframed within a concern for people's livelihoods and expectations that rehabilitated agricultural systems can both create employment and produce food.

Information is a component within such livelihood support that has often been overlooked in the rush to supply material assets such as food, seeds or infrastructure. Even when aid agencies recognize the need for 'soft' inputs such as information, the supply-driven dynamics of their work still lean toward efforts to 'do livelihoods' by handing out predetermined assets – cash or in kind (Christoplos 2006, 2011a).

The tendency to overlook knowledge in favour of providing packages is also rooted in a simple notion that recovery is about merely returning to what people did before. Plans for contributions to livelihoods and food security are often anchored in nostalgia for a pastoral vision of the past. For example, many programmes assume that refugees want to return to their supposed rural roots even though they have spent years (or even generations) in the semi-urban environment of refugee camps or in the slums of neighbouring countries. They may never have been farmers – and may not have ever intended to become farmers – but nostalgia creates an illusion that a hoe and a bag of seeds is all they need to 'pick up where they left off'.

This is often paired with hopes that it is both possible and desirable to rebuild the state-led institutional structures of the past, be they large public-sector extension agencies or cooperatives directed by the government. Civil servants and politicians who have been fighting in the bush may return to the capital with a remembrance of the system that was in place decades earlier. For better or for worse, their new government probably cannot generate revenues to cover the recurrent cost of a new army of extension agents.

Agricultural recovery programmes are inevitably selective in terms of whose livelihoods are supported and tend to search for 'beneficiaries' that match the package they want to provide. For example, claims are frequently made that seed programmes target the 'most vulnerable groups' in places where the most vulnerable people are, for the most part, landless and have nowhere to plant the seeds they receive. In many recovery efforts relatively large investments are channelled to entrepreneurs and farmers who have the land, knowledge and social capital to take advantage of prevailing opportunities (Christoplos 2007, 2011a, Christoplos et al. 2010). These programmes 'pick winners' with assumptions that this will lead to a trickle-down effect that creates livelihoods for the chronic poor. There has been insufficient research into whether or not these assumptions are valid, but there are indications that 'picking winners' may instead ignore large numbers of losers in recovery.

This is an example of an area where agricultural advisory services will inevitably be self-selected by a limited group of clients/beneficiaries. Information is often asymmetrical, as technologies or market knowledge that are relevant for a 'winner' may be irrelevant to a poorer farmer, producers that live far from markets, or those with no choice but to farm ecosystems where risks of floods, droughts or rapid environmental degradation are high and willingness to invest for the future is therefore low. These 'losers' are usually subsistence-oriented and may even be more interested in information about how to get out of farming and do something else.

Can Advice to Micro-farmers Lead to Future Livelihoods in Rural Bosnia-Herzegovina?[1]

Bosnia-Herzegovina (BiH) is a country where the choices of whom to support with agricultural rehabilitation have been difficult. Even if it is apparent that aid to 'winners' is unlikely to trickle down, the prospects for significant and sustainable impacts of agricultural rehabilitation that is directed to the rural poor are not promising either. Such efforts cannot reverse overall trends toward a concentration of agriculture in larger commercial farms. This has led to a context where tough choices are needed as there is no ideal information package that will lead to development of economically sustainable farming enterprises and also meet the needs of the rural poor.

In order to understand the choices of rural people and aid providers it is important to analyse the way post-conflict recovery emerged. In the immediate post-war period the main reconstruction priority in BiH was neither agriculture nor livelihoods – it was housing. Massive programmes were initiated with dual objectives of providing permanent shelter and also of reversing ethnic cleansing. The destruction of housing (primarily owned by the Muslim or 'Bosniak' population) during the war was used as a way to try to ensure that people would not be able to return to their former homes. The international community was determined to reverse this ethnic cleansing by assisting people to return to what was assumed to be their home villages. Agricultural programming was initially introduced as a complement to housing programmes as it was increasingly recognized that there were few livelihood opportunities available to the returnees in rural areas. Over time, the need to find employment for returnees turned into an ever more pressing issue as people registered to receive rural housing and promptly left for urban areas and for Western Europe. In 2005 20 per cent of the housing constructed by the international community after the war was estimated to be vacant (Skotte 2005). This was because many of those who received houses had actually already moved to urban areas before the war or emigrated during the war. They had rural roots but had lost their links to rural livelihoods years (or even generations) before. Many rural houses today are effectively used as holiday homes by families that have settled in other countries. The moral imperative of reversing ethnic cleansing came to be combined with the additional political imperative of discouraging emigration by supporting BiH refugees to return to rural 'homes' (and stay there).

One outcome of this combination of factors was the emergence of a development narrative in BiH which assumed that agriculture would generate new jobs. The need to find a solution that would keep people in their rebuilt homes created pressures to look rather optimistically on the potential of addressing Bosnia's employment crisis through farming. This was in direct contrast to the rest of Eastern Europe where policies generally acknowledge that future trends will involve a shift to

---

1   This case study is based on a study conducted for the Overseas Development Institute (Christoplos 2007).

larger mechanized production units and declining job opportunities in farming itself. Livelihoods programming in BiH tried to move in the opposite direction.

Post-war agricultural rehabilitation programming was polarized between two approaches. The first type of programme grew out of reintegration efforts. They developed from the 'livelihood components' that were originally established as a relatively minor complement to housing reconstruction efforts. The main target group was the unemployed. The large majority of the beneficiaries of these programmes farmed considerably less than one hectare and did not farm at all before receiving assistance. Attention to the long-term viability of the micro-farming enterprises being supported was often secondary to the need to support livelihoods in the short term. These programmes originally provided seeds and tools and gradually developed further into multicomponent initiatives combining very intensive technical advisory services, cooperative development, provision of small greenhouses, credit, and assistance in marketing.

The second set of programmes focused more exclusively on modalities of promoting growth in agricultural production, productivity and profitability. These programmes focused on 'picking winners' and provided them with generous assistance. The types of services provided were in many respects similar to those provided for micro-farmers, but the modalities were much more closely linked to strengthening knowledge in relation to changing market conditions and value chains. The 'winners' supported by these programmes were usually relatively well-off farmers with larger landholdings. The link to poverty alleviation and reintegration in these programmes was sometimes tenuous, but there was an assumption that a well-functioning and competitive agricultural sector would be at least somewhat more 'pro-poor' than a large number of semi-subsistence farmers with few prospects of reaching markets.

In both sets of programming the link between objectives of reintegration and growth was usually presumed to be in the development of farmers' associations and cooperatives to provide advice for their members and to market agricultural products. There was an assumption (or at least a hope) that farmers' groups would take over the role of aid-financed NGOs and manage sustainable market linkages for the poor. There were reasons to be cautious in accepting this assumption at face value. Although strong cooperatives certainly have the potential of generating employment for the poor in processing, transport, and so on, and perhaps reducing the costs of some services, there was no sign that they were ready to welcome the poorest producers as active and equal members. Even if they were to try, it was unlikely that farmers' associations could develop capacities that would absorb the very high transaction costs of providing information and advice for diverse and scattered micro-farmers. It was doubtful that the heavy investments by the NGOs in reaching the poor would be covered in the future by the wealthier farmers in the cooperatives.

In order to understand the prospects for agricultural rehabilitation to prove sustainable, it is important to look at the market for the crops being produced. BiH has followed the advice of the international community in terms of agreeing to open its markets to trade and globalization. As part of this process BiH is in the midst of an exponential shift in its agrifood systems. Supermarkets and shopping centres are being constructed throughout the country, even in relatively small towns. In 2006 BiH had 19 hypermarkets (with over 12 cash registers) and 250 supermarkets (with two to 11 cash registers) (Huisenga 2006). The drastic changes in supply chains and consumer preferences that are occurring will have profound impact on both the risks and opportunities facing farmers in BiH. If they can meet the demands of supermarkets for quality, quantity and timeliness at a competitive price, their market opportunities are enormous. If they cannot, they may lose their markets entirely. The demand through small shops and open markets that smallholders supplied in the past is declining. The micro-farmers who received post-war support have to dramatically increase the quality and quantity of their production if they are to retain access to the commercial markets of the future. Their prospects are not good.

In partial recognition of such challenges, agricultural programming in BiH gradually became increasingly *marketing-oriented* in the sense that NGOs engaged directly in helping producers sell their crops. Some adopted virtual contract farming arrangements with their 'beneficiaries', providing them with inputs and buying their crops for resale to retail outlets. This does not mean that these programmes were necessarily *market-oriented*, in terms of being based on supporting farmers' own understanding of markets, competitiveness and profitability. In interviews, some private-sector actors who bought strawberries and raspberries from NGO-supported and -organized farmers complained that the NGOs with which they worked showed remarkably little concern about strengthening their clients' capacities to maintain production levels and profits. This was said to be because the NGOs rely on a donor (rather than the market itself) for their own funding, and were therefore not genuinely accountable for the ultimate viability of the farming enterprises they supported. From the perspective of these private-sector critics, such accountability can only emerge if the NGOs' operational survival was to be related to their results in terms of production and agricultural profitability.

There were some positive examples of successful micro-farmers who have taken advantage of agricultural rehabilitation support to scale up their production, and some of these will certainly survive. There were model farmers who used agricultural rehabilitation support as a springboard into competitive market agriculture. It is, however, doubtful that the majority of those who received this aid will be able to earn a significant income within the changing agrifood systems in BiH.

Seed and Sustainable Extension in Afghanistan[2]

After 2001 a large proportion of agricultural rehabilitation programming in Afghanistan focused on seed distributions. This was justified by three assumptions.

---

2   This case study is based on studies conducted for the Afghanistan Research and Evaluation Unit (Christoplos 2004) and the Overseas Development Institute (Longley,

First, it was assumed that there was an absolute lack of seeds due to the conflict. Second, seed was seen as a way to subsidize and thereby contribute to recapitalizing agricultural production, thereby accelerating recovery. Third, there was a perceived need to increase access to 'improved' seed and to speed up genetic renewal.

Success in relation to this third technology-transfer objective has been exceptional – in interviews, agency staff proudly declared that in some areas local varieties had disappeared entirely, a tendency that is disturbing for those who see value in retaining Afghanistan's extraordinary agrobiodiversity. The rapid rate of adoption of new varieties of technology indicates that Afghan farmers were not satisfied with the seeds they had access to. A central question was how technology-transfer efforts could be promoted in a manner that recognizes the rights of farmers to know what they are getting involved in when presented with an 'improved' technology. These rights were not being upheld in the immediate post-2001 period. In 2002 some private European companies were supported by their national governments to distribute seed varieties that had not been tested in Afghanistan. The production results were appallingly poor. This naturally increased the food insecurity of the farmers who planted the seeds and had an extremely negative impact on the credibility of the agencies that provided and recommended these seeds.

An additional question regarding the choice of instruments for agricultural rehabilitation concerns whether the menu of interventions was congruent with the nature of the challenges that rural people were facing in re-establishing their livelihoods. Even where far broader needs were ostensibly recognized, including information, seed often formed the bulk of items received by farmers. This stemmed from three factors:

- There was a legacy of humanitarian and early rehabilitation programming, wherein seed was perceived to be the only feasible form of intervention (among many NGOs)
- Aid agencies adapted their ambition levels to the institutional weaknesses that made other types of activities hard to establish. Seeds were relatively easy to distribute without strong local partners
- Agencies had difficulties in accessing longer-term funding for other types of initiatives since Afghanistan was then still assumed to be in a 'humanitarian phase'.

These factors, common in many post-conflict contexts, resulted in what has been referred to elsewhere as the 'seeds and tools treadmill' (Remington et al. 2002). But Afghan farmers wanted more than seeds and tools. After 2001 Afghan agricultural stakeholders optimistically expected the state 'to provide everything!!' (Hemani 2003). This was in contrast to many development agencies that believed that the weakness of Afghan institutions represented an opportunity for public reform

Christoplos and Slaymaker 2006).

through creation of a leaner state and privatization of a range of functions, a paradigm often referred to as New Public Management (Christoplos 2000, 2004). This hope that a smaller, more efficient public role can emerge from crisis is built more on prevailing narratives within the aid community than on local expectations and empirical realities. Afghan politicians and civil servants had a different set of priorities and incentives, based on a different history and culture from that of their advisers. They faced a different set of political, economic and social pressures. With the fall of the Taliban the vast majority of politicians and civil servants in agriculture took for granted that a large and well-financed state-led agricultural development effort was going to be put into place. That was what they had experienced under past regimes (both during the Soviet occupation and during earlier periods of US dominance), and there was little understanding or awareness of the public-sector reform efforts that had been undertaken in many other countries over the past two decades. New Public Management was not embraced at local levels either. All of those interviewed in provincial and district political and civil service structures retained expectations of a return to state-led development, requiring a large bureaucracy and public provision of extension services.

Nonetheless, under pressure from the international community the Afghan government eventually accepted to limit its role in agriculture to regulation, leaving NGOs, consulting firms contracted by aid donors and Provincial Reconstruction Teams (led by the international military forces) to provide services instead, with these agencies claiming spectacular results (Koch, Harder and Saisi 2010). Sometimes foreign military forces are themselves directly providing agricultural advisory services and are still distributing seeds (ReliefWeb 2010, 2011). In recent years there have been a number of private-sector-led initiatives that are similar to the approaches of 'picking winners' projects in BiH (see http://afghanistan.usaid. gov/en/USAID/Activity/1/Accelerating_Sustainable_Agriculture_Program_ ASAP). Observers have noted that faith in market-led agricultural growth is not shared by the rural poor, who tend to distrust risky markets and choose to pursue improved livelihoods by searching for non-farm employment while they continue with subsistence agriculture (Kantor and Pain 2011). When the poor distrust markets, projects that 'pick winners' are likely to be self-selected by those farmers with better market connections and conditions.

The reform narrative behind these approaches has its roots in efforts around the world to reduce the role and scale of state bureaucracies to regulatory and facilitation functions, implying sizeable reductions in public-sector staff. A paradox in Afghanistan was that there was little agricultural bureaucracy to actually retrench. Instead there was a need to expand staffing, even if it was just to meet the challenges of regulation. There was a contradiction between the retrenchment narrative and the actual make-up of the agricultural civil service. Even though the state eventually abandoned its hopes for engaging again in direct service provision, the challenges it faced in shouldering the regulatory and facilitation tasks envisaged in the reform narrative were enormous. Border controls, inspection of imported fertilizer and pesticides and the need to monitor and regulate a transparent system for agricultural finance were all pressing needs where capacity would need to be enhanced.

The need to rebuild these basic functions of the state was not just a concern for the international community. Wide-open borders allowed traders to try to sell anything, with few deterrents (apart from losing future customers) for selling bogus products. When asked about information needs, farmers interviewed were most vehement in insisting that they needed to know whether the agricultural inputs they were buying were really fertilizer or pesticide or if they were just white powder. The information they wanted was related to their presumably well-justified distrust of the free market and their search for ways to deal with the negative consequences of deregulation and a weak state.

Advice on Efficient Water Use in Central Asia[3]

The Ferghana Valley in Central Asia is one of the main hotspots of transboundary water conflict. Kyrgyzstan, Tajikistan and Uzbekistan meet in a fertile and densely populated valley (seven million inhabitants) with a jigsaw of borders, rivers, irrigation canals and transport networks. During the Soviet period, this complicated system created interdependencies that provided a basis for cooperation among the three republics. Today the same system generates tensions. Most of the water flowing into the valley originates in the poor mountainous areas of Kyrgyzstan before flowing into the other two republics. Tajikistan is the driest area, and is therefore planning to build a dam to ensure that it controls sufficient water for its farmers. Uzbekistan, which is considerably better off than its two neighbours, lies downstream and has responded to Tajikistan's plans with anger. While Uzbekistan remains a strong state, poverty and chronic conflict have resulted in declining state capacities in Tajikistan and Kyrgyzstan, which the International Crisis Group (ICG) has described as being 'haunted by the increasingly likely prospect of catastrophic systemic collapse' (2011).

The distribution of different ethnic groups does not follow national borders. In 2010, ethnic fighting in Kyrgyzstan resulted in massacres and destruction of homes and businesses of ethnic Uzbeks in the Kyrgyz city of Osh, with a significant proportion of the Uzbeks fleeing. Borders were closed and tensions increased. Earlier recurrent ethnic conflict in Tajikistan has now subsided, but for many years these areas of the Ferghana Valley were on the 'seeds and tools treadmill'.

One way to reduce tensions over water is to increase the efficiency with which the water that does reach farmers' fields is actually used. Since 2008 the Swiss Agency for Development Cooperation has been supporting a small regional project, the Water Productivity Improvement at Plot Level Project (WPI-PL), in

---

3   This case study is based on findings from a mid-term review (Christoplos 2011b) of the Water Productivity Improvement at Plot Level Project, financed by the Swiss Agency for Development Cooperation.

the three countries to develop a system for advice to farmers on how to measure and more efficiently use irrigation water. Initial results show a potential for water savings of up to 50 per cent with increased yields (WPI-PL 2010). If the farmers need less water, then the pressures on existing supplies will decrease, potentially decreasing political and social tensions as well.

Naturally, things are not so simple. The very different social, political and aid systems in the three countries have resulted in three very different systems emerging for providing this advice, each with its strengths and weaknesses.

In Kyrgyzstan WPI-PL is 'competing' for farmers' attention with humanitarian agencies offering generous free inputs and services and therefore has found it necessary to provide a more lavish package of support than in Tajikistan or Uzbekistan. Osh is still a base for significant humanitarian operations and the post-conflict Kyrgyz state is still struggling to determine its role in agricultural development. Years of development and humanitarian support have resulted in the existence of a number of relatively strong private/NGO service providers, skilled in providing services (including advice on more efficient water use) to farmers. But they have also resulted in very weak ownership and engagement in these services from central and local government. Farmers have learnt that their best opportunity to access advice is through aid projects and the various NGOs and private firms that implement them.

Uzbekistan is at the opposite end of the spectrum. NGOs are non-existent and the project therefore works entirely through public-sector institutions and local water user associations. In general, aid agencies in Uzbekistan have no choice but to work through the government, even if they have concerns about where governance is headed (ICG 2011). The regime in Uzbekistan recognizes that it is dependent on cotton and wheat from the Ferghana Valley, and is determined to retain control over production. In order to maximize production they are providing incentives and directives to create larger farms, while requiring the shrinking number of farmers to provide jobs for a set number of labourers. The state maintains a virtual contract farming relationship with farmers. The government instructs farmers what to plant, provides them with seeds, fertilizers and other inputs and buys their harvests at controlled prices. Water is provided free, which would seem to be a disincentive for efficient water use. But the limited range of production factors that farmers control has meant that they are particularly eager to find ways that they themselves can improve yields, and more closely matching crop requirements with irrigation is one of the few production factors that they can still control.

Until a few years ago the aid reaching the Ferghana Valley in Tajikistan was primarily humanitarian, with seeds a major component. Now these projects have been phased out and the state is looking for new ways to mobilize water user associations to take a leading role in production decisions. Compared to its two neighbours, Tajikistan has chosen a more market-driven approach to water use. Farmers pay a much higher price for water than in Kyrgyzstan and Uzbekistan, and are thus very interested in minimizing these costs. Water user associations are eager to learn how to optimize their use of water and also to find ways to convince local authorities to shift to charging for the water that they actually use, rather than a flat rate per hectare, which has been the standard procedure until now.

Knowledge about how to optimize water use, both in relation to crops and in terms of the institutions required (for example, water user associations), is potentially a strategic factor in managing conflict in Central Asia. The experience of promoting more efficient water use shows that success is possible, but that the institutional structures that are needed to facilitate such technological change will vary according to the prevailing political economy and legacy of both humanitarian programming and the strong statist approaches of the Soviet era. Even in countries with similar histories, very different institutional systems and social contracts can emerge. Kyrgyzstan has developed into a weak state where farmers are used to relying on ad hoc aid projects for accessing information services. Uzbekistan has retained many accoutrements of an anachronistic Soviet system, with farmers accustomed to receiving virtual instructions rather than advice. Tajikistan experienced a long period when public support for agricultural services was heavily influenced by international humanitarian actors, with consequent uncertainty about what rural development norms should replace the departing humanitarians. Farmers in the three countries thus have very different access to information and different expectations of their service providers.

Conclusions

After a conflict there are inevitably new demands, challenges and opportunities for agricultural advisory service development. There are four aspects that need to be considered:

Breaking Out of Nostalgic Mindsets

States and citizens usually enter into recovery with hopes that the structures of the past can be rebuilt. In Afghanistan there were expectations that the grand irrigation schemes, parastatal marketing boards, state-managed cooperatives, large public advisory service agencies and other accoutrements of the statist models of economic development would be revived. A desire to rebuild a long-lost rural economy in order to reverse ethnic cleansing and discourage emigration encouraged international and national NGOs in BiH to promote projects that were disconnected from current market realities. It is important to not only recognize that the assumptions behind choices in agricultural recovery were 'wrong', but also that there are historical trajectories that generate these outmoded development narratives.

Such nostalgia may overshadow understanding (by both aid providers and recipients) of what the post-conflict state can do to provide agricultural services and what the implications are if they cannot (or do not wish to) shoulder these tasks. The gap between the desire to do 'everything' and limited capacities to do very much leads some countries, such as Kyrgyzstan, to eschew responsibilities.

States that have avoided becoming 'fragile', such as Uzbekistan, are wary of the examples of their neighbours and resist reform, clinging to strong statist models of the past as long as they can afford them.

Confronting Difficult Choices in 'Picking Winners' and Encouraging New Livelihoods

These examples also show that nostalgia exists alongside modernization narratives that point in the opposite direction. 'Picking winners' is based on acceptance of the notion that many of the rural poor are better off doing something else. In Uzbekistan policies specifically encourage concentration of landownership. Some of the recovery investments in BiH and Afghanistan indirectly favour farmers with the resources and knowledge required to maintain relatively large, commercial farms. They require specialized information services, which will be more likely to be provided by private actors, often within contract farming arrangements.

Reassessing What the State Can Do and What It Must Do to Rebuild a Social Contract

In development efforts recovery processes are often portrayed as an opportunity to use investments to introduce New Public Management. Indeed, conflicts do drain the capacities of civil service agencies and severely restrict tax collection. Whether these conditions are seen to represent opportunities or obstacles relates to the ideology of the observer. There is no obvious solution. In post-conflict contexts the demand for publicly supported agricultural services is extremely high. Ability to provide front-line and visible services is perceived by the public as evidence of whether or not the state is back in operation. Public expectations in Afghanistan and BiH in many respects reflect a slowly fading hope for a return of the paternalistic state that existed in the past. State-led extension services were provided in the past in all the case-study countries, but lack of public revenues makes a return to these systems (and continuation of the prevailing system in Uzbekistan) unlikely. Alternatives based on pluralistic mobilization of a range of service providers (Christoplos 2010) can only be built if the foundations of a market for these services exist. These service providers need to be paid for by the state, aid agencies, private investors or their clients. In the aftermath of a war there are usually very few local agricultural advisory service providers (either NGOs or private firms) ready to bid for contracts. There is also generally little or no capacity within government structures to be the 'smart buyer' of services that New Public Management demands. Experience in Kyrgyzstan and Tajikistan shows that such a market for extension services can be developed in time. But it cannot be created overnight and being aid-financed, its sustainability is uncertain.

The advisory service systems that emerge in post-conflict contexts are rarely models for pluralistic, multistakeholder collaboration. Neither do they resemble the large public bureaucracies of the past. Instead complex combinations often emerge wherein a jumble of public, private and civil society structures jostle for funds and clients. The hybrid and multiple institutions that evolve are patchy and do not provide a uniform array of services, either geographically or in terms of types of services. They are rarely accountable to either their clients or the government as they are usually busy hustling their next aid contract.

In some post-conflict contexts the state is happy to delegate agricultural advisory services to 'somebody else' as they are overstretched trying to maintain other services or reconstructing infrastructure. Advice to farmers is often seen as generating less political 'value for money' than investments in infrastructure. In Kyrgyzstan and Tajikistan the government water authorities are primarily concerned with investing available resources in repairing and maintaining irrigation canals. They are happy if somebody else can give advice to farmers about how to use the water they receive. In other countries relations between NGOs and state providers of advisory services are marked by mutual distrust. NGOs are accountable to donors instead of state structures, which can reinforce citizens' perceptions that their political leaders at national and local levels are weak and not responsive to their needs (Vaux and Visman 2005, Berry et al. 2004). This is particularly true for front-line services, such as agricultural advice.

From Silver Bullets to Institutional Transformations

Humanitarian agencies usually 'tick the sustainability box' in service provision by claiming that they will 'hand over' service provision structures to local government, farmer groups or 'the community'. In practice, local authorities generally cannot or will not live up to their part of the deal. Farmer organizations can rarely be formed within the short time frames of humanitarian programmes and are likely to be subject to the same inter-ethnic tensions that led to the conflict. 'Communities' turn out to be an assortment of individuals with a range of interests. Handovers tend to result in unstaffed offices or agents without an operational budget to leave their office and meet with farmers.

Although some institutions may collapse during war, various formal and informal institutions continue to provide services. Amid seeming collapse there is usually something to build upon. An understanding of what services exist and how farmers are accessing information can shed a different light on how to move forward without the ideological blinkers of New Public Management or nostalgia for bureaucracies of the past. The everyday realities of war are not all-encompassing, and jumping to assumptions about war 'risks disabling precisely the strategies and tools of social organization, culture and politics through which violence can be reduced and its adverse effects mitigated' (Richards 2005). The alternative is to recognize that services are needed, but even with the best of services many people will not want to go back to farming, nor will they be able to develop the competencies needed to compete in today's agrifood systems (Weiss Fagen 2011).

References

Berry, C., Forter, A., Sultan, S. and Moreno-Torres, M. 2004. *Approaches to Improving the Delivery of Social Services in Difficult Environments*. PRDE Working Paper 3. London: DFID.

Christoplos, I. 2000. Natural disasters, complex emergencies and public services: rejuxtaposing the narratives after Hurricane Mitch, in *Applying Public Administration in Development: Guideposts to the Future*, edited by P. Collins. Chichester: John Wiley & Sons, 345–58.

Christoplos, I. 2004. *Out of Step? Agricultural Policy and Afghan Livelihoods*. Kabul: Afghanistan Research and Evaluation Unit.

Christoplos, I. 2006. *Links Between Relief, Rehabilitation and Development in the Tsunami Response*. London: Tsunami Evaluation Coalition.

Christoplos, I. 2007. *Between the CAPs: Agricultural Policies, Programming and the Market in Bosnia and Herzegovina*. HPG Background Paper. London: ODI.

Christoplos, I. 2010. *Mobilizing the Potential of Rural and Agricultural Extension*. Rome: FAO and the Global Forum for Rural Advisory Services.

Christoplos, I. 2011a. Food security and disaster, in *Routledge Handbook of Natural Hazards and Disaster Risk Reduction and Management*, edited by B. Wisner, J.C. Gaillard and I. Kelman. Oxon: Routledge.

Christoplos, I. 2011b. External review of the Water Productivity Improvement at Plot Level Project (WPI). Unpublished.

Christoplos, I., Rodríguez, T., Schipper, L., Narvaez, E.A., Bayres Mejia, K.M., Buitrago, R., Gómez, L. and Pérez, F.J. 2010. Learning from recovery after Hurricane Mitch. *Disasters*, 34 (issue supplement s2), s202–s219.

Hemani, M. 2003. *Agriculture and Rural Development Ministrie's [sic] Level Strategy Development Workshop, 28 September–2 October 2003*. Kabul: Ministry of Rural Rehabilitation and Development.

Huisenga, M. 2006. Supermarket Report, LAMP, USAID Sarajevo (unpublished).

International Crisis Group (ICG). 2011. *Central Asia: Decay and Decline*. Asia Report 201. ICG.

Kantor, P. and Pain, A. 2011. *Running Out of Options: Tracing Rural Afghan Livelihoods*. Afghanistan Research and Evaluation Unit Synthesis Paper Series.

Koch, T.K., Harder, A. and Saisi, P. 2010. The provision of extension services in Afghanistan: what is happening. *Journal of International Agricultural and Extension Education*, 17(1), 5–12.

Longley, C., Christoplos, I. and Slaymaker, T. 2006. *Agricultural Rehabilitation: Mapping the Linkages Between Humanitarian Relief, Social Protection and Development*. HPG Research Report 22. London: Overseas Development Institute.

ReliefWeb. 2010. Afghanistan: agribusiness development team provides farming classes for local farmers in Khost province [Online]. Available at: http://reliefweb.int/node/357359 [accessed: 20 June 2011].

ReliefWeb. 2011. Afghanistan: British troops teach animal care in Helmand [Online]. Available at: http://reliefweb.int/node/391149 [accessed: 20 June 2011].

Remington, T., Walsh, S., Charles, E., Maroko, J. and Omanga, P. 2002. Getting off the seeds and tools treadmill with CRS seed vouchers and fairs. *Disasters*, 26(4), 316–28.

Richards, P. (ed.). 2005. *No Peace No War: An Anthropology of Contemporary Armed Conflict*. Oxford: James Curry.

Skotte, H. 2005. Rebuilding the community of Grapska, in *Returning Home: An Evaluation of Sida's Integrated Area Programmes in Bosnia and Herzegovina*. Sida Evaluation 05/18. Stockholm: Sida.

Vaux, T. and Visman, E. 2005. *Service Provision in Countries Emerging from Conflict*. Bradford: Centre for International Cooperation and Security, University of Bradford.

Water Productivity Improvement at Plot Level (WPI-PL). 2010. *Innovative Partnership on the Way to Water Productivity Improvement*. Tashkent: Water Productivity Improvement at Plot Level.

Weiss Fagen, P. 2011. *Refugees and IDPs after Conflict: Why They Do Not Go Home*. Special Report 268. Washington, DC: United States Institute of Peace.

Chapter 16

Youth, Associations and Urban Food Security in Post-war Sierra Leone

Roy Maconachie

Introduction

Sierra Leone has recently emerged from a long period of instability and dislocation following the end of a protracted decade-long civil war during the 1990s. The country's economy and quality of life deteriorated rapidly during this period, when many people were forced to flee their homes and abandon their mainly agricultural livelihoods due to the rebel insurgency. As a consequence, the population of the country's capital city, Freetown, almost doubled in size, with youths being at the forefront of this influx. As agricultural production in rural areas became severely dislocated, food security and the state of the agricultural sector in the post-conflict period have become a major concern for governmental and non-governmental development agencies.

While early efforts in post-war reconstruction focused primarily on rural development and the rehabilitation of rural livelihoods (see UN 2004), more recently, attention has shifted to a daunting new set of problems unfolding in urban areas. A variety of economic and human development indicators suggest that Sierra Leone continues to be ranked as one of the world's poorest countries (UNDP 2009), and concerns amongst international donors and government policymakers about rising levels of urban unemployment and food insecurity have returned to centre stage. These concerns are often framed against the backdrop of a series of challenges associated with the social reintegration of marginalized young men and women into society, and the underlying 'crisis of youth' narrative that continues to drive post-war development discourse and policymaking.[1] Indeed, the challenges faced by the country's youth cohort are acute, if merely for the fact

---

1   The 'crisis of youth' narrative has particular salience for increasing pressures associated with urban areas in Sierra Leone. In a number of informal settlements in Freetown, particularly in the east and central parts of the city, reports suggest that there are high concentrations of ex-combatants, many of whom did not formally undergo the disarmament, demobilization and reintegration (DDR) process and lack employment skills and training. This remains a source of great concern in Sierra Leonean government and donor circles. Indeed, 15–30-year-old males are the very demographic group that is most likely to resume warfare if left frustrated and excluded (MSI 2004).

that they represent such a significant proportion of the population. According to the 2004 census, 79 per cent of Sierra Leone's total population is under the age of 35 years and 33 per cent is between the ages of 15 and 35. Nearly 70 per cent of the latter group is unemployed and 53 per cent are illiterate.

In the aftermath of the war, the government's focus on both job creation and food security was heavily reflected in the 'second pillar' of Sierra Leone's first Poverty Reduction Strategy Paper, developed for the 2005–7 period. In the second PRSP of 2009 (Government of Sierra Leone 2009) these concerns were once again prioritized as key issues, but particularly for the country's youth population. While there continues to be much concern about the lack of opportunity and uncertain future for young people in Sierra Leone, this may be particularly so for those living in urban areas. One of the demographic consequences of forced displacement during the civil war was that the country's population became much more mobile and more urban (Brown et al. 2005). The capital city, Freetown, received a large proportion of an estimated 2 million internally displaced persons (IDPs) who sought both protection and livelihood support in the city (IDMC 2004). It is estimated that more than 500,000 farm families were displaced by the conflict, and agricultural production was so severely dislocated that by 2001 only 20 per cent of the annual rice requirement (the staple food) was produced in the country (EIU 2002).

During the war, not only did food production in rural areas become severely dislocated, but Freetown became cut off from its rural food supply catchment. This was especially serious during the interregnum period in 1997, when there was a complete land, sea and air blockade in the Western Area surrounding Freetown. Since this time, the urban food supply has been put under further pressure from a continuing stream of rural-to-urban migration. Commentators suggest that, contrary to orthodox explanations for the 'urban bias' phenomenon, the reasons for migration into Freetown extend beyond mere economic motivations. For example, it has been pointed out that some youth 'may find opportunities for coexistence, reinvention and empowerment in cities' (Sommers 2007: 8). Alternatively, others have suggested that former child combatants are one group that has a particular interest in living in cities, both because they are afraid to return to their rural villages, and because they consider rural life to be undesirable (Peters, Richards and Vlassenroot 2003: 28). Recent statistics suggest that 44 per cent of Sierra Leone's population now lives in urban areas, rising to 53 per cent among 20–24-year-olds and 58 per cent among 25–29-year-olds (UN 2006).

Since 2007, social problems in cities have been compounded by the global spike in the cost of food and rising oil prices, which have made urban life even more challenging for Freetown's residents. Estimates suggest that the cost of rice has risen by 300 per cent, and the issue of urban food security has gained increasing importance on policy agendas. According to a recent FAO/WFP report:

> Market prices in Freetown and inflation data, as well as qualitative evidence, indicate a severe crisis of food security in urban areas. For example, in Freetown

the price of a bag of rice is approximately Le 125,000,[2] or 83% of the Le 150,000 minimum wage. With escalating food prices (even for non-globally traded goods), stagnant income and high unemployment, we may expect purchasing power to decline in the urban areas. Urban populations, therefore, constitute the most immediate crisis-affected areas. [FAO/WFP 2008: 6–7]

Indeed, evidence from other cities in sub-Saharan Africa suggests that both the sustainability and political stability of burgeoning African cities will be crucially dependent upon the availability of affordable food supplies for growing populations (ADB 2009, Walton and Seddon 1994). Under conditions of post-war economic hardship, rising food prices and a reported decline in patronage from 'big men' in Sierra Leone, urban youth have been one of the most affected groups in the country and they have had to engage in new forms of social cooperation and develop new livelihood strategies to make ends meet. A recent World Bank study of youth employment reports that the post-war period has seen 'a remarkable upsurge in self-organized social activism among young people' (Peeters et al. 2009: 17), and a parallel revival of reciprocal labour groups, cooperative associations and rotational credit clubs has been noted in both rural and urban areas countrywide (Fanthorpe 2007).

This chapter focuses on the question of urban food security in and around Freetown and explores an increasingly important response to the rising demand for food in the city: a proliferation of urban and peri-urban agriculture (UPA). While it is apparent that UPA has become a vital strategy for combating the high cost of food and generating a significant amount of household income, the chapter also highlights how UPA activities may be currently driving resurgence in community-based associations amongst young people, a development that could have important implications for shaping wider change in the post-conflict period.

Drawing upon field-based research carried out in Freetown between 2007 and 2010, the chapter explores the knowledge and perceptions of urban farmers who are involved in Freetown's grassroots UPA associations, with a particular interest in youth participation. While the age distribution of those involved in UPA reflects a broad cross-section of society – both young and old – evidence suggests that contrary to the widespread belief that youth today are not interested in agriculture (see Government of Sierra Leone 2009: 98), there is, in fact, significant participation by young people in UPA activities. Although other researchers have suggested that many youths in Sierra Leone are 'disinterested in dedicating themselves to agriculture' (Peters, Richards and Vlassenroot 2003: 28), this research indicates that 62 per cent of UPA farmers surveyed were under the age of 35 years, while 27 per cent were under the age of 15 years (see Figure 16.1). Moreover, while evidence suggests that the motivation for youth participation in UPA associations appears to vary considerably, the discussion demonstrates that

---

2   In July 2010, US$1 was equivalent to *c*.3,850 Leones (Le). So at the time, one bag of rice in Freetown cost approximately US$32.46.

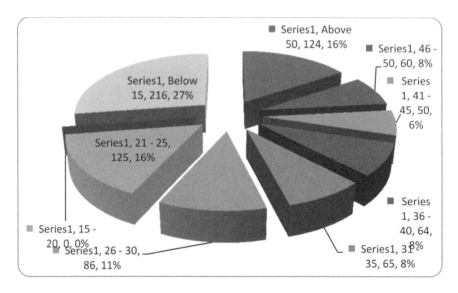

Figure 16.1   Age distribution of urban farmers

Source: Figures based on author's survey administered to 483 respondents in 2009.

these associations serve as an important mechanism to strengthen their livelihood portfolios and supplement household food supplies, while allowing young people to diversify their strategies and gain access to development resources in new ways. In particular, some dynamic youth leaders have realized that by forming UPA associations – many of which serve as umbrella organizations for a wide range of other sub-associations – they are able to 'capture' NGO and government support that they would not normally have access to.

Urban and Peri-urban Agriculture in Freetown

Whilst the post-war proliferation of UPA activities in and around Freetown has recently caught the attention of government policymakers, food production in the city is not a new phenomenon. Indeed, UPA has been practised in Freetown for many years and has long served as a key source of household nutrition, income and employment. While UPA made important contributions to urban food security in Freetown through the early years of independence, and particularly during the difficult period of the one-party state under President Siaka Stevens (1971–85), it later emerged as a critical factor in food and livelihood security for both rural refugees and urban residents during the civil war. However, it was not until after the end of the conflict that UPA was officially recognized by policymakers and donors as a major contributor to Freetown's urban food supply. Most notably,

Sierra Leone's World Bank/FAO-sponsored Agriculture Sector Review (MAFFS 2004) acknowledged UPA as a vital livelihood activity for alleviating poverty and boosting urban food supply, while at the same time stressing the need to support urban agricultural associations. Since then, a number of UPA programmes have been initiated in Freetown by local and international NGOs, including the International Water Management Institute (IWMI), the Dutch group Resource Centres on Urban Agriculture and Food Security (RUAF) and the Italian NGO Cooperazione Internazionale (COOPI).

Over the years, it has thus been apparent that UPA in Freetown has been a major contributor to urban food security, helping to satisfy growing consumer demand for fresh foodstuffs in a rapidly growing city and also providing valuable financial rewards for producers and sellers. The research reveals that farming activities currently involve the production of 32 different crops, and include ten species of animals in domestication and husbandry activities. The most commonly cultivated crops are 'exotic' vegetables (cabbage, lettuce, carrots, spring onions, tomatoes and beans) and locally consumed leafy vegetable crops (potato leaves, spinach and cassava leaves). Because these crops are perishable and cannot withstand long-haul transportation, most are consumed locally and on a daily basis. They are usually harvested and sold at the market on the same day.

While the positive aspects of UPA have been well recognized by both policymakers and development practitioners alike, in the case of Freetown little field-based research has been carried out on the dynamics of UPA production, or the strategies undertaken by farmers to maximize their benefits. Moreover, there has been no research into how UPA associations relate to aspects of youth livelihoods in a post-conflict context. The fieldwork upon which this chapter is based was carried out in a series of stages between 2007 and 2010 and focuses on these issues, particularly as they relate to the role that UPA plays in post-conflict food security and income generation.

In January 2008, Google Earth images of Freetown and the Western Area were analysed and 59 UPA sites were identified in greater Freetown. These sites were located on a 1:50,000 topographic map sheet of the Western Area, and the information was cross-checked at each location to ensure that it was accurate. During the following six-month period, baseline information on the present status of UPA was collected through a combination of questionnaire surveys administered to a sample of 483 urban farmers and semi-structured interviews and focus group discussions at each of the sites. Over the following year, further detailed information was obtained through focus group discussions and interviews with key informants, including data on the role that agricultural associations are currently playing in UPA production and marketing. Finally, between May and July 2010, all 59 UPA sites were visited and 63 farmer associations were identified to be operating. Detailed focus group discussions were held with a third of these associations (21 associations in total), ensuring that there was equal geographic representation from all wards in Freetown (see Figure 16.2). Many of the respondents were young people, and were able to provide detailed information

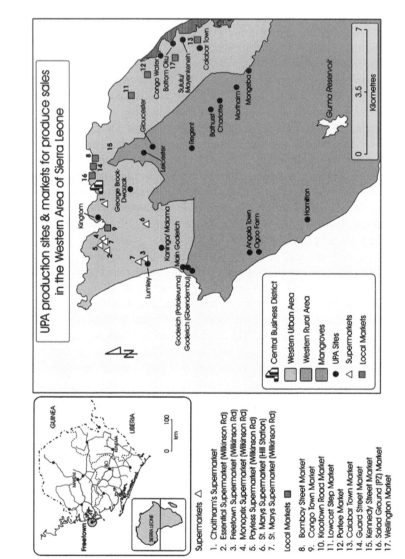

Figure 16.2 Sampling sites for discussions with UPA associations

Source: Author's fieldwork.

as to how UPA associations were addressing the problems of urban youth. The discussion that follows examines some of the key findings of this fieldwork.

Associations, Youth and UPA

In Sierra Leone, it has been suggested that 'horizontally organized' interest-driven associations have taken on a new salience in the country's period of post-war recovery. Elizabeth Watson points out that dynamic, resilient associations, and the institutions that cement them, can play a fundamental role in mitigating livelihood vulnerability often associated with conflict and population displacement (Watson 2001). However, in societies that have been seriously affected by violent conflict, such as Sierra Leone, there are often complex challenges in rebuilding social capital, which remains vital in operationalizing collective action and mobilizing community-driven development (Richards, Bah and Vincent 2004). Of critical significance is the way in which social networks and relationships of reciprocity that exist between individuals are reshaped by more powerful actors, which can in turn affect the character and strength of community-based institutions.

Post-war research carried out in the Eastern Province of the country in Kono district suggests that the renewed interest that many young people are presently showing towards farming and family-based associational life may be helping to drive a resurgence in community-based cooperation in rural areas, an important foundation stone for creating a more durable form of democratic change (see Fanthorpe and Maconachie 2010). However, attention to examples of self-organized, interest-driven associations in other parts of the country suggests that an assessment of social change in the context of post-war aid intervention is often extremely difficult to gauge. For example, the now disbanded Agricultural Business Unit (ABU) initiative was one unsuccessful donor attempt to restructure farming associations and mobilize the labour power of young men and women across the country. The experience demonstrated that local interests in rural society are embedded in unequal social relationships in far-reaching ways, and ultimately there is a danger that decentralized agricultural programmes such as the ABU initiative may become a new source of patrimonial resources for the elite (see Maconachie 2008).

In the context of Freetown, our research revealed that many urban farmers had memberships in local associations, and that self-organized groups were playing an increasingly important role in both the production and marketing of foodstuffs. While the size and make-up of these groups varied considerably, the focus group discussions with 21 UPA associations in July 2010 revealed that the majority of farmers were from poor socio-economic backgrounds, with over 80 per cent lacking formal sector employment. The research also revealed that the majority of the 63 associations identified were operating mainly in the east end and West Ward III of Freetown. The size of individual associations varied significantly, ranging from a membership of just ten farmers to as many as 168, with the average size of an association being 32 farmers. The majority of urban farmers also indicated that they were migrants from the provinces, most of whom had moved to Freetown during the war. Our survey revealed that the dominant ethnic groups involved in UPA are Loko (27 per cent) and Limba (27 per cent), followed by Temne (24 per cent) and Mende (13 per cent). Creoles represent only 1 per cent of urban farmers in Freetown.

The fieldwork also suggested that there are multiple challenges and motivations for people's involvement in UPA associational life. Broadly, focus group discussions revealed that members believed that associations were necessary to meet a variety of constraints associated with urban food production in the post-conflict era. Referring to Table 16.1, where a subset of the most popular reasons for forming an association is presented, it is apparent that there are a number of common factors that motivate urban farmers to participate in associational life. Most typically, three kinds of production constraints were noted: (i) problems of gaining access to urban land to cultivate, (ii) difficulties in mobilizing farm labour, and (iii) challenges in getting foodstuffs to the market to sell. Many young people believed that by coming together as an association, they could begin to address these challenges together. In the discussion to follow, each of these constraints will be explored in turn.

(i) Land Insecurity

A baseline survey carried out for this study in 2009 demonstrated that more than 31 per cent of respondents interviewed believed that the primary constraint that they faced concerned their limited access to land and insecure tenure arrangements in urban and peri-urban areas. Although this may be a constraint that has particular relevance to young people, who often lack the networks and resources required to gain access to land, land hunger for urban plots was generally a problem that was reported by respondents of all ages.

As Table 16.2 illustrates, UPA farmers are able to gain access to land in a variety of ways, but the majority of farmers interviewed reported that they were either renting their urban plots from private landowners, or had made informal arrangements with municipal authorities to cultivate small parcels of 'idle' government land. In both cases, cultivators complained that they were vulnerable to high levels of exploitation, particularly as tenure arrangements were informal and not guided by well-defined laws or policies. However, by participating in an association, many young farmers believed that this gave them more collective bargaining power in terms of gaining access to land from 'big men'.

In urban areas where available land is particularly scarce, the relationship forged between the association and the private landowner appears to be vital. In some cases this relationship can be mutually beneficial, but in other instances there is potential for conflict to occur, particularly over disagreements on the terms of tenure and the amount of remuneration given in exchange for access to land. In a number of instances, association members reported that they had taken to building additional alliances with other 'big men' as a way of protecting themselves. Although

Table 16.1 Factors that motivate participation in UPA associations: summary of statements from focus groups in Freetown

Reason for forming association	% of associations that referred to reason (n=21)
Pooling labour makes it easier to generate income (employment through vegetable sales)	71
With weekly contributions in the association we have been able to start a standing fund	66
The association has opened up opportunities to solicit financial support from donors/government	62
As a group it is easier to deal with labour constraints	57
As an association, land access/tenure is more secure	52
The association serves as an umbrella for other social activities, which are important for resocializing youths	48
The association helps us to access tools/inputs from agricultural extension agents	43
As a group we can more effectively deal with marketing and transportation constraints	38
The association serves as a good mechanism to diversify our income-generating activities	38
As a group we can produce more food for household consumption	33
As an association we have more bargaining power with local authorities	33
The association helps us to access knowledge transfer from extension agents	29
We can work together to address environmental issues	24

Source: Author's field data: July 2010.

Table 16.2   Access to UPA land amongst respondents (n=483)

Landownership	%
Self-owned	9.7
Family-owned	7.9
Community-owned	9.1
Rented	42.1
Government land	27.6
Other	3.7

Source: Author's fieldwork.

most groups tended to avoid affiliation with political parties, it was noted by a number of them that their association was very dependent on establishing strong relationships with those more powerful individuals who could assist them, particularly in terms of helping them to secure their land tenure. The situation is explained by one young farmer from the Angola Youth Farmers Association:

> Normally, we go into negotiation [with a landowner]. The rent involves two payments, one before using the land which is usually about Le 50,000 [*c.* US$13] and then one after using the land, which involves giving a portion of the harvest to the landowner. But it is also sometimes good to meet with the local authorities, such as the section chief or headman, as well. Local authorities can serve as further security in case the landowner denies the agreement later on. [personal communication, 9 July 2010]

In a number of cases, interviews also revealed that private landowners have now actually begun to encourage associations to work their 'idle' land. Indeed, local reports of land grabbing in Freetown are becoming much more common,[3] and some landowners reported fears that their vacant land would be taken over by urban businessmen. Many were thus very happy to have an association cultivating their land. In the words of the Chairwoman of the New England Vegetable Growers' Association:

---

3   Numerous reports have recently appeared in Freetown's newspapers about the increased incidence of land grabbing. For example, see Jalloh 2010.

When you leave your land empty without putting up any structures, there is a high possibility that people will come and grab your land. If somebody important sees that nobody is using a plot of land, particularly if it is in a strategic place, they may use their influence to occupy it forcefully or through their political connections. But if you have an association that is constantly cultivating the land, that will make it more secure. That is a kind of guarantee for the landowner. [personal communication, 12 July 2010]

In sum, land hunger and insecure tenure arrangements are serious constraints for all UPA farmers in Freetown, particularly as there is currently a lack of well-developed policies in place to regulate the use and management of municipal land for agriculture. This problem has been acknowledged by both government and international development donors, and in response the EU is supporting the preparation of a new 'Freetown Master Development Plan', which should help to address the land question. Similarly, the recent decision by the Ministry of Lands to move away from a contract-based tenure system to a title-based land tenure system in the Western Area should make it easier to identify, protect and allocate more suitable land to urban farmers (RUAF 2010). FUPAP (the Freetown Urban and Peri-urban Agriculture Platform) is also presently engaged in brokering an agreement between the Ministry of Lands, Ministry of Agriculture, Forestry and Food Security (MAFFS) and the local authorities, so that appropriate land for UPA can be identified by the government and then allocated through long-term users' agreements to associations registered with MAFFS.

(ii) Labour Constraints

Perhaps more significantly, the association members interviewed for this study spoke of an urgent need to address the issue of labour. Elsewhere, other commentators have suggested that traditional rotational labour clubs, which have a long history throughout the country, have become much more important in the post-war recovery process, and have served as the basis for reviving self-organized associational life (Richards, Bah and Vincent 2004). Such communal labour clubs have existed for many years, and traditionally have involved the coordination of large reciprocal work gangs (referred to as *boma* groups in Kono and *kombi* groups in Mende) that move from farm to farm to spread the burden of work over the farming season. These labour inputs from young men and women have been vital in undertaking a sequence of demanding farm activities that must take place within a brief window of opportunity in the agricultural cycle. However, to activate such labour cycles, certain resources are required (that is, the ability to provide 'food for work'). The desperate material poverty that many farmers currently face in the post-conflict period has made this challenging and has also meant that they are often 'labour-poor' (see Maconachie 2008).

Resurgence in associational life has thus been an important element in rejuvenating farming systems, particularly in the immediate aftermath of the war when a reduction in the availability of family farm labour and increased levels of poverty made agriculture extremely challenging. Research carried out elsewhere in Sierra Leone has demonstrated that considerable benefits accrue to those farmers who possess the networks, social skills and opportunities to organize large working parties (Maconachie 2008). But as the range of redistributive processes that occur within communities have been greatly challenged in the post-war period, the impacts on reciprocal labour relationships have been dramatic. In many ways, the labour functions that associations provide have helped to fill this gap.

In addition to reciprocal tasking on farms, the ability of UPA associations to mobilize groups of young, able-bodied labourers appears to contribute to a range of other important activities as well. For example, the Wanword Farmers' Association is an organization of 46 farmers who cultivate in the Congo Water region in the eastern part of Freetown. The farm plots that members cultivate are situated on the margins of a coastal mangrove swamp and plots are regularly inundated by salt water, which destroys the crops. The original reason why the association was formed, members explained, was to meet the heavy labour demands required to build a bund along the periphery of the farmland to keep out the sea water. In particular, the mobilization of a group of strong, able-bodied young men was required to undertake this physically demanding and labour-intensive task. But from this initial labour need, the association evolved to take on a variety of other interests and functions, many of which served to raise the financial capital required to maintain the bund.

Consequently, an additional function of many UPA associations, it was noted, is that they can be hired out to provide labour, an activity which generates cash income. With the income generated, many groups, such as Wanword Farmers' Association, have also set up 'standing funds' to assist members with emergency expenses. The injection of this income into association activities has many positive livelihood spin-offs. For example, many young women in the associations interviewed reported that a portion of income generated by the association was also often used to engage in *osusu*, a kind of local revolving microcredit scheme.

(iii) Marketing Challenges

A final constraint noted in focus group discussions with UPA associations concerned the marketing challenges encountered in the sale of urban produce to consumers. These constraints resonate clearly with recent academic interest in how urban food supply systems are shaped by the role of urban markets, particularly in how they regulate the agricultural produce trade (see Porter et al. (2004) for a good summary of the literature). Indeed, a number of NGOs have also argued that adding value to urban agriculture produce through food processing and marketing is an important way of generating income and new employment opportunities in developing countries. In short, by linking food production, processing and marketing, urban farmers can potentially generate higher and more regular returns for their produce, a condition that could reduce the social exclusion of the urban poor, and particularly unemployed youth (IDRC 2003).

In this study, a significant constraint reported by respondents concerned poor access to urban markets. Many urban farmers interviewed noted that they often had no choice but to sell their produce to 'middlemen' at much lower farm gate prices, as they lacked the ability to access markets themselves. Several farmers pointed out that this situation created little economic incentive to increase production, since it was likely that any surplus produce would not arrive at the central markets while it was still fresh, and would likely perish on their plots. While researchers elsewhere have used classical locational analysis of farming to suggest that high-value and perishable crops should be produced in close proximity to the urban market (Ellis and Sumberg 1998), in Freetown the 'remoteness' of markets often encompasses more than just physical distance from the point of production. A lack of affordable transportation and poor relationships with urban buyers were also reported as being major constraints in getting produce to the market.

Focus group interviews revealed that 38 per cent of the associations in the sample believed that they could more effectively deal with marketing and transportation constraints as a group than they could individually. Lack of transportation was reported as being one of the major barriers to getting access to markets and, according to the baseline survey, 75 per cent of UPA producers reported that they were resorting to head-loading their produce to urban markets. In some cases, it was revealed that 'standing fund' contributions from members of the association were being used to hire group transportation to take produce to the market.

Thus, many association members believed that the marketing challenges faced by urban farmers could be greatly ameliorated if appropriate government interventions took place. According to one member of the Lower Gbendembu Farmers' Association:

> A major area of intervention that we would like the city council to help us with is to provide a regulated space to market our produce. Usually when we take our crops to the market there are problems with sellers who have already established themselves. The sellers are very protective of their spaces and there is no opportunity for us. Sometimes our produce is thrown away. The only alternative is for us to carry our crops to other places to sell. This is a major problem because even if we produce good-quality vegetables, there is still the issue of selling them to the buyer. [personal communication, 13 July 2010]

While a number of key institutions, such as the Ministry of Agriculture, Forestry and Food Security (MAFFS), the National Association of Farmers of Sierra Leone (NAFSL) and Freetown City Council (FCC), have begun providing agricultural extension services to UPA associations, there has so far been relatively little attention to the marketing constraints that farmers face. The EU-funded FUPAP project aims to stimulate innovation in urban agriculture in Freetown through support to UPA associations, emerging commercial producers and to youth interested in agricultural production and processing. Moreover, various stakeholders coordinated under FUPAP have subsequently started to work with groups of vulnerable youth on commercial agricultural activities in Freetown, including value chain analysis and business development, group strengthening and participatory life-skills training ranging from communication, leadership and decision-making to conflict management and literacy and numeracy (RUAF 2010). While these are all important developments, more attention must now be given to facilitating improvements to the urban agriculture marketing chain, a development which could serve as an important pathway for providing employment for youth.

Conclusion

While there is now a significant body of research exploring the livelihood benefits of UPA in sub-Saharan Africa, there has been remarkably little academic enquiry into the role that cooperative associations assume in facilitating the activities and processes involved in urban food production. In the context of Sierra Leone, such associations are playing an increasingly important role in addressing the issue of urban food security during the post-conflict period, during which large numbers of young people, many of them ex-combatants, are actively involved in urban farming.

At the same time, particularly during a period of rising global food prices, there is widespread concern in Sierra Leone that the existence of a mass of unemployed and disaffected youth could readily provide the catalyst for a return to instability and conflict. Nearly 70 per cent of 15–35-year-olds are jobless, contributing to one of the highest unemployment rates in the world and further perpetuating fears associated with a 'crisis of youth'. As the Africa Economic Development Institute (AEDI) recently noted, 'With so many Sierra Leonean youths unemployed, there is a potential threat of an amalgamation of youth's disaffection, chronic poverty and high cost of living to the stability of the nation' (AEDI 2010). These fears were confirmed at a meeting of the UN Security Council in March 2010, which expressed concern that, 'despite the magnitude and political significance of the country's youth unemployment problem, relatively little progress had been made in addressing it' (UN 2010). Although the seriousness of the youth unemployment situation has been recognized by the Sierra Leone government, and in December 2009 a Youth Commission Act was passed in parliament, it remains to be seen whether this will lead to any tangible progress in incorporating youths into both economy and society.

Given these significant challenges faced by youth, the involvement of young people in productive cooperative associations is a particularly positive development. The spontaneous organization of self-help associations, in some cases incorporating traditional practices of rotating credit and communal work, is an encouraging trend in an environment where a resource-deprived government has a long agenda, not least in the massive task of rehabilitating war-damaged infrastructure. Evidence suggests that such associations not only play a key role in rebuilding the social capital networks that were destroyed during the war, but are also helping to safeguard urban food security. With their greater bargaining power, producer associations can mitigate major constraints associated with urban food production, such as acquiring land for growing crops, providing farm labour and gaining better access to marketing opportunities. Both the municipal authorities and NGOs need to recognize these constraints, engage in dialogue and work towards supporting producer associations.

Whereas in some African cities there are varying degrees of hostility towards the practice of UPA, it seems that the Freetown city authorities are adopting a rather more progressive approach and are carefully considering the position of UPA in future sustainable urban planning strategies. But in terms of the broader economic development framework in such a poor post-conflict country, the building of trust and cooperation as witnessed in UPA activities must be encouraged and where possible more widely replicated. Indeed, the resurgence in community-based associations amongst young people is a development that could have important implications for shaping wider change in the post-war period. Surely there must be lessons here for other post-conflict African states and beyond.

Acknowledgements

The research upon which this chapter is based was carried out in collaboration with Fourah Bay College, Sierra Leone, under a British Council/DFID-funded DelPHE project. The author would like to thank Kabba Bangura for his invaluable assistance in the field.

References

African Development Bank (ADB). 2009. *Soaring Food Prices and Africa's Vulnerability and Responses: An Update*. Working Paper No. 97. Tunis: African Development Bank Group [Online]. Available at: http:/www.afdb.org/ [accessed: 1 June 2011].

Africa Economic Development Institute (AEDI). 2010. Sierra Leone youth unemployment. AEDI [Online]. Available at: http://www.africaecon.org/ index.php/africa_business_reports/read/53 [accessed: 1 June 2011].

Brown, T., Fanthorpe, R., Gardener, J., Gberie, L. and Gibril Sesay, M. 2005. *Sierra Leone: Drivers of Change*. Bristol: IDL Group.

Economist Intelligence Unit (EIU). 2002. *Country Profile: Guinea, Sierra Leone, Liberia*. London: EIU.

Ellis, F. and Sumberg, J. 1998. Food production, urban areas and policy responses. *World Development*, 26(2), 213–25.

Fanthorpe, R. 2007. BRACE (Building Resilience and Community Engagement) Institutional Survey. Concern Worldwide (unpublished).

Fanthorpe, R. and Maconachie, R. 2010. Beyond the 'crisis of youth'? Mining, farming and civil society in post-war Sierra Leone. *African Affairs*, 109/435, 251–72.

Food and Agriculture Organization (FAO) and World Food Programme (WFP). 2008. *The Severe Impact of Rising Food Prices: A Situational Assessment of the Food Crisis in Sierra Leone*. Rome: FAO and WFP.

Government of Sierra Leone. 2009. *An Agenda for Change: Second Poverty Reduction Strategy (PRSP II), 2008–2012*. Freetown: Government of Sierra Leone.

Internal Displacement Monitoring Centre (IDMC). 2004. Sierra Leone: executive summary. IDMC [Online]. Available at: http://www.internal-displacement. org/8025708F004CE90B/(httpCountrySummaries)/0372C6E093AFEFEB80 2570C00056B6D0?OpenDocument&count=10000 [accessed: 1 June 2011].

International Development Research Centre (IDRC). 2003. Guidelines for municipal policymaking on urban agriculture, No. 9: Processing and marketing urban agriculture products. IDRC [Online]. Available at: http://www.idrc.ca/ uploads/user-S/10530126310E9.pdf [accessed: 1 June 2011].

Jalloh, A.R. 2010. Rampant land grabbing in Koya. *Concord Times*, 20 September [Online]. Available at: http://allafrica.com/stories/201009210192.html [accessed: 1 June 2011].

Maconachie, R. 2008. New agricultural frontiers in post-conflict Sierra Leone? Exploring institutional challenges for wetland management in the Eastern Province. *Journal of Modern African Studies*, 46(2), 235–66.

Management Systems International (MSI). 2004. *Integrated Diamond Management in Sierra Leone: A Two-year Pilot Project*. Washington, DC: USAID.

Ministry of Agriculture, Forestry and Food Security (MAFFS). 2004. *Agriculture Sector Review*. Freetown: MAFFS.

Peeters, P., Cunningham, W., Acharya, G. and Van Adams, A. 2009. *Youth Employment in Sierra Leone: Sustainable Livelihood Opportunities in a Post-Conflict Setting*. Washington, DC: World Bank.

Peters, K., Richards, P. and Vlassenroot, P. 2003. What happens to youth after conflict? A preliminary scan of the Africa-related literature and assessment of debates. Den Haag: Netherlands Development Assistance Research Council.

Porter, G., Lyon, F., Potts, D. and Bowyer-Bower, T. 2004. *Improving Market Institutions and Urban Food Supplies for the Urban Poor: A Comparative Study of Nigeria and Zambia*. Wallingford: Department for International Development [Online]. Available at: http://eprints.mdx.ac.uk/3818/1/ Improving_market_institutions_and_urban_food_supplies.pdf [accessed: 1 June 2011].

Resource Centres on Urban Agriculture and Food Security (RUAF). 2010. Freetown (Sierra Leone). RUAF [Online]. Available at: http://www.ruaf.org/ node/1133 [accessed: 1 June 2011].

Richards, P., Bah, K. and Vincent, J. 2004. *Social Capital and Survival: Prospects for Community-driven Development in Post-conflict Sierra Leone*. Social Development Paper No. 12. Washington, DC: World Bank.

Sommers, M. 2007. *West Africa's Youth Employment Challenge: The Case of Guinea, Liberia, Sierra Leone and Côte d'Ivoire*. Vienna: United Nations Industrial Development Organization.

UN. 2004. *United Nations Transitional Appeal for Relief and Recovery: Sierra Leone 2004*. Freetown: United Nations.

UN. 2006. *UN Support to Sierra Leone National Youth Employment Programme (NYEP)*. Freetown: United Nations.

UN. 2010. Youth unemployment poses 'latent threat' to Sierra Leone's stability. Security Council 6291st Meeting. UN [Online]. Available at: http://www.un.org/News/Press/docs/2010/sc9890.doc.htm [accessed: 1 June 2011].

UNDP. 2009. *Human Development Report, 2009*. Basingstoke: Palgrave Macmillan.

Walton, J. and Seddon, D. 1994. *Free Markets and Food Riots*. Oxford: Blackwell.

Watson, E. 2001. *Inter-institutional Alliances and Conflicts in Natural Resource Management*. Marena Working Paper 4. Cambridge: Department of Geography, University of Cambridge.

Chapter 17

The Only Way to Produce Food is to Cooperate and Reconcile? Failures of Cooperative Agriculture in Post-war Sierra Leone

Catherine Bolten

Introduction

The civil war in Sierra Leone between 1991 and 2001 had a devastating effect on agriculture. The fighting, which involved the Revolutionary United Front (RUF) rebels, government forces and local civil militias, forced many farmers to flee their villages.[1] They returned home, often years later, to find their tools stolen or broken and their seed stocks looted. Fields that had lain fallow for a decade required intensive labour – typically provided by young men – to clear. This labour was generally unavailable as many youths who resided as displaced persons in urban areas during the war remained to pursue employment and education opportunities. Unable to clear the seasonally flooded *boli* land, most farmers engaged instead in small-scale production of secondary crops. For farming villages located in proximity to disarmament centres, funding for large-scale production of staples was only available if civilian farmers agreed to form cooperatives with ex-combatants. Ex-combatants – mainly the demobilized fighters of the RUF – were a primary 'target beneficiary group' of post-war aid projects because of the prevalent fear that an unhappy ex-combatant would return to living by the gun (see Utas 2005). Thus, unlike civilian non-combatants, they had ready access to agricultural aid.

In 2002, the northern district of Bombali was home to more than 4,000 ex-combatants, most of whom looked to aid as a potential route to economic self-sufficiency. They were largely from the south and lacked the social connections that would provide access to jobs and arable land. These ex-combatants hesitated to return south without the means to support their families, and cooperative farming undertaken in conjunction with willing civilian villages was one of the few possibilities for income. Funding was available specifically for cooperative agricultural projects between ex-combatants and civilian farmers. Donors

---

1   Unless otherwise stated, the author gathered all of the information in this chapter during fieldwork between February and July 2005, and between May and July 2010.

promoted cooperative-agriculture-as-reconciliation because it resonated with the discourse of 'coexistence' popular at the time (see Chayes and Minow 2003: 7), even as the lack of alternatives to cooperation meant, essentially, that it was the *only* option for agricultural funding. Ex-combatants formed community-based organizations (CBOs), and actively solicited villages to cooperate with them, as this was the only way ex-combatants could gain access to grants, land and labour. Civilians, who lacked recourse to funding for food production outside of cooperation, often agreed.

In this chapter, I argue that cooperative agriculture must be voluntary, rather than obligatory, if ventures between ex-combatants and civilians are to thrive. Based on a CBO called PADO whose activities I observed between 2004 and 2010, I illuminate how civilian resentment of being forced to cooperate with ex-combatants – whom they already viewed as benefiting unfairly from post-war aid – ultimately resulted in the project's failure. Their civilian partners in the village of Malempa were so desperate to attract donor funding that they contributed their personal land to a communal farm that would largely benefit the ex-combatant participants. This caused friction that was exacerbated by ex-combatants assuming ownership of the project and treating the civilian farmers as their labour pool. For their part, ex-combatants balked at civilian demands that 'cooperation' meant that ex-combatants would provide either labour or wages for civilians' work on the communal plot. A civilian who wanted revenge on ex-combatants eventually stole the rice harvest, thus revealing the fault lines of cooperation. As a result, I suggest that food security be treated separately from reconciliation programmes in post-war aid. Though cooperative agriculture can assist reconciliation efforts, it should not be conducted in isolation from other efforts to restart agriculture along the lines desired by civilians, as this defeats the voluntary nature of cooperation by which reconciliation is enabled. In this case, farmers wanted individual control of individual plots from which they could prosper for themselves. Before the project began, they had lacked enough rice to feed their families for five months, and had no ability to create or sell a surplus that would enable the payment of their children's school fees. The primary goal was self-determination; reconciliation activities were a means to that end.

DDR Imagines Agriculture as a Reconciliation Activity, Rather than Food Production

The National Commission for Disarmament, Demobilization and Reintegration (NCDDR) was established in 2001 to prepare ex-combatants for civilian life and create opportunities for reconciliation. As the war had lasted for over a decade, DDR architects believed that the bulk of ex-combatants – most of whom were under the age of thirty – had very little memory or understanding of civilian life or how to relate to civilians outside war.[2] NCDDR undertook two interrelated

---

2   Interview with NCDDR programme manager, Freetown, 6 July 2003.

programmes to address ex-combatants' need to reintegrate: skills training and cooperative agriculture.

Combatants were initially brought into demobilization camps, where they were disarmed, given provisions and a disarmament allowance, and registered for skills training. Ex-combatants were offered six months of training in the skill of their choice, transition allowances to ease their reintegration into the community, and a final travel allowance to encourage them to return home. Many problems existed with DDR, the most important being shoddy training in skills in which master artisans had already cornered the market, which contributed to ex-combatants' inability to find work. They turned instead to the second component of the programme, in which a bilateral donor offered start-up money for cooperative agricultural CBOs.

According to the director of one NGO tasked with initiating these programmes who I interviewed in June 2003, 'The idea is to engage ex-combatants and civilians in farming, cooking and eating together, so that they remember that each of them is a person.' Designed for reconciliation rather than food production, his organization's programmes aimed to tap into a fundamental Sierra Leonean understanding of positive personhood: the necessity of productive, nurturing cooperative activity. The director believed these initiatives would begin the healing process, and there was no emphasis on the success of actual agricultural production listed among his NGO's project goals.

The development NGOs operating in Bombali district after the war targeted villages they considered 'rural' – more than twenty miles from the nearest market town – for agricultural assistance. Villages located near the district capital of Makeni were thus ruled out as 'not vulnerable' (see Bolten 2008: 336). Donors targeted these nearby villages for reconciliation activities instead, for the very reason that their proximity to Makeni ensured the presence of ex-combatants, who congregated in the urban areas. Having been denied funding by development NGOs and donors, villages such as Malempa agreed to work with ex-combatant-organized CBOs under the rubric of reconciliation.

PADO Implements Food Crop Production for Sustainability with Civilian Labour

Four former members of the RUF registered the Polong Agricultural Development Organization (PADO) in 2002, having solicited a participation agreement from the village of Malempa.[3] The stated goal of the organization was to 'alleviate the hunger and starvation that are a major component of poverty by farming cooperatively'. PADO solicited funding for the project Food Crop Production for Sustainability in 2004, and the proposal emphasized introducing mechanization as a way of easing the burden of land clearance, which historically was undertaken by men. The initial grant requested seed stock, hand tools, tractors, concrete

---

3   All persons, villages and organizations have been given pseudonyms.

platforms on which to dry newly harvested rice and a new building for rice and seed stock storage. It described the provision of rice seed and tools to individual plots as well as the creation of a communal seed bank that would provide income for the organization:

> PADO plans to work with community people [farmers] towards the development of agriculture on individual and communal bases. Planting areas have been selected. With funds available we will procure and distribute tools to beneficiaries to prepare the land. The individual farms are for members to be self-sufficient in what they grow. The communal farms are to serve as a seed bank and to raise funds for the organization. The beneficiaries are 560 community people and 38 ex-combatants.

The grant, written by the ex-combatants, perhaps unconsciously emphasized the ex-combatants' pre-eminence as the primary organizers of PADO – 'PADO plans to work with community people' – essentially stating that *they* were PADO, and that the communal farm would therefore benefit them, as the 'organization'. Five civilian landowners donated the land for the communal plot, from which they received no individual benefit. The organization would provide seeds, tools and labour to clear individual and collective farms, and civilians and 34 ex-combatants (not including the four PADO board members) would work the communal farm to grow seed stock for the individual (civilian) and collective farms, with the surplus sold to support PADO, essentially, to support the ex-combatants. As was required by the donor, the proposal outlined plans for 'improved reintegration of ex-combatants into agricultural communities' through regular reconciliation workshops. £5,000 was budgeted to pay for transportation, food and expenses to convene workshops with a trained facilitator. The stated purpose of these workshops was conversation and communal consumption, and thus the reintegration of ex-combatants in the community.

The division of labour was made clear in the proposal. The 'beneficiaries' – civilians in Malempa – would undertake the labour on their own plots and the communal one, and reap the harvest of their own plots. Ex-combatants would work only the communal plot, the single largest plot in the project. The ex-combatants would gain income from the outputs of the communal plot, on which civilians provided the bulk of the labour. It was unclear both from the proposal and implementation what labour PADO board members provided aside from supervision and the initial work to obtain the grant. The board members made sure that civilians had the seed, land clearance and tools to work their own plots; however, if civilians were able to secure funding for their own agricultural projects, they would not be required to donate and farm a communal plot for ex-combatants. This became a contentious issue, as I will describe later.

In 2005, the project received a £90,000 grant from a bilateral donor. The donor stipulated that the project must be supervised by an experienced national NGO, so PADO partnered with Our Future, headquartered in Makeni, to manage the budget and logistics. Our Future established a bank account on behalf of PADO, and was tasked with purchasing materials and seed, supervising construction of drying floors and rice mills, and organizing reconciliation workshops. Our Future was expected to visit Malempa regularly to monitor productivity in addition to their work completing project infrastructure.

Controlling the Budget: Agriculture Is about Self-determination, Not Cooperation

Project implementation was marked by arguments over funds between the board members of PADO and the director of Our Future, and between PADO and the civilian beneficiaries. PADO complained that Our Future was mishandling funds because procurement and implementation were slow. Our Future argued that the bank delayed the funds and construction materials and seed were scarce, and civilian farmers demanded a share of the budget in cash. Farmers balked at working on the communal plot without being paid by PADO for their labour, despite the fact that the project funded seed, tools and clearance for individual plots. They argued that ex-combatants must contribute more and not treat the project as a favour to the community from which they themselves benefited, as would chiefs ordering serfs to farm their plots.[4]

The argument that PADO should pay for civilian labour on the communal plot also stemmed from unresolved reintegration resentment. Ex-combatants who registered with NCDDR – thus rendering them eligible to create CBOs – received a cash allowance for initial disarmament, one for travel, one for skills training registration and once a month during the six months of training. The lavishness of these benefits caused 'grumbling' among civilians, who received nothing during reintegration and complained that ex-combatants were the primary beneficiaries of both war and peace (see Ginifer 2003: 46). Residents of Malempa wanted ex-combatants to share these benefits, thus making a gesture to redress the wrongs of the war and the unfairness of reintegration. PADO board members, claiming that they were redressing the wrongs of the war by engaging in the project in good faith, refused. They blamed the lack of official reconciliation activities on Our Future's failure to implement workshops, and reminded civilians that the bulk of future seed stock would come from the communal plot. The ex-combatants acted as supervisors, managers and project owners, and not as equal participants, which angered civilians. At one meeting in Malempa when the women's representative asked for wages, the board president replied, 'But we have given you this project!' He understood the ex-combatant 'labour' being the work they put into the grant proposal and making connections with bilateral donors. The women halted work on the communal plot for a month in protest, and worked only on their individual plots.

---

4   Under the colonial administration, chiefs in the northern district abused their right to tributary labour, which was one of the grievances sparking a region-wide rebellion in 1955 (see Fanthorpe 2001: 380).

The arguments between PADO and Our Future stemmed from the linkage between budget control and project control. Money and materials were so scarce that everyone wanted right of use and disbursement. The director of Our Future argued that as he had been given responsibility for disbursement of funds, he also had the right to use project materials for Our Future's other ventures. The ex-combatants documented implementation failures attributable to Our Future out of concern for a donor audit because the project was proceeding so slowly. In correspondence with the director, PADO complained about the unwillingness of the NGO to sign a memorandum of understanding with the CBO. They cited the failure of Our Future to purchase cement for the construction of drying floors and storage sheds, lack of reconciliation meetings, the unwillingness of Our Future to provide money to train civilians in building maintenance and the fact that furniture and supplies for PADO's office were never purchased as delays that set the project back a year. Our Future, it appeared, did not want to take orders from an ex-combatant CBO.

Arguments over the budget may seem commonplace in agricultural initiatives, but in this case they highlighted much deeper issues than quibbling over pennies. Each set of actors wanted to ensure that funds were utilized to their own best interests – not cooperative best interests – illuminating the enforced nature of the cooperation and individuals' emphasis on maximizing their ability to produce and control their own outputs. Civilians and Our Future resented having to cooperate with and take orders from PADO, which they viewed as taking advantage of them in order to produce their own income with minimal effort.

Confrontation Instead of Reconciliation: Seizing Control of Agricultural Outputs

The project proceeded in 2005 when Our Future produced two tractors to till individual and communal plots, fuel for the tractors, and hired drivers to plough the tough *boli* land. The NGO procured 40 bushels of seed rice for individual farms and 20 bushels for the collective farm. The 34 ex-combatants worked on the collective farm and exhorted civilians to join them by emphasizing the need for a seed stock for the next year's planting. Despite the lack of drying floors or a rice mill, the project produced 270 bushels of rice on the communal farm in its first season.

One ex-combatant remembered the palpable joy as squabbling and lack of infrastructure were put aside to produce a bountiful harvest:

> The community [Malempa farmers] were so happy, they were working so well together when they saw that we would have a great season that they didn't think too much about everything that the project was lacking. There was no drying floor so we improvised, taking the rice to the highway to dry it along the verges. We still had this problem of storing the rice, since there was no mill building, but

we pulled together again and did a home-made building, mud brick and thatch, that was ready to store the seed for the next year.

He recounted the prevailing optimism as Malempa farmers and the PADO board met to decide collectively on rice distribution. It was only once these meetings began that the head of Our Future became an active participant:

> This man started coming to the community again once the rice was harvested. He was very interested in what we were going to do with the rice, now that the programme was successful, despite the fact that he did not offer us assistance or finance aside from the use of the tractors and the seed rice. We as a community organization decided on how much rice would go to which purpose: some for members to eat, some for each family to sell and some for us to reserve as seed for the group to plant in the next season. We all agreed to this.

Some rice was also earmarked for PADO, and he noted that Malempa residents did not protest because in the initial proposal, *all* of the rice from the communal farm was allocated for seed and PADO's own support. The PADO board was learning that the project's success hinged on them being responsive to their civilian collaborators, but this was not enough.

The following day, the director of Our Future led a convoy of vehicles into Malempa. He told residents that Our Future was following through on PADO's decision to take the rice to town in order to weigh it and to do administrative work required by the donor. Thus appeased, residents watched while the rice was loaded into the vehicles. The next day the rice was not returned as the director had promised, and Our Future never returned to the village. The director disposed of the rice himself.

Enforced Cooperation Was the Node around which Resentment Turned

Before describing the reactions of the residents of Malempa to the theft of the rice, it is critical to understand *resentment*, which will illuminate cause and consequences of the theft. Resentment can be triggered by any perceived injustice. According to the philosopher Jeffrie Murphy, 'resentment stand[s] as emotional testimony that we care about ourselves and our rights' (2003: 19). He argues that resentment is not akin to vengeance, but a desire for acknowledgement that wrongs have been committed and require redress. If consistently unaddressed, resentment can spiral into desire for revenge.

In this case, enforced cooperation intensified existing grievances civilians had with ex-combatants, as well as creating new grievances. They resented being forced to undertake reconciliation activities because they were not eligible for agricultural funding on their own merits as a farming community. Their resentment of the ex-combatants themselves – the patronizing attitude they held towards their civilian labourers – exacerbated cooperation problems, thus hindering civilians' will and ability to concentrate on food production. There were no reconciliation meetings, whether or not a facilitator could be engaged, for unattended wrongs from war or reintegration to be aired, and PADO resisted being flexible with their project to satisfy their collaborators. A token wage for farming the communal plot would have helped redress both the unfairness of ex-combatant reintegration and the unfairness of enforced cooperation. The paltry sums requested by the farmers – the equivalent of £0.40 per person per day – constituted a symbolic levelling. It required only that PADO acknowledge that civilians were their equals, and were not providing free labour in exchange for the opportunity to farm their own plots. The fact that this never occurred triggered among Malempa residents a resentful reaction, a 'shrink[ing of] the world down to a tightly defensive constrictive coil' (Solomon 1983: 51). The possibility for self-determination was mocked, thus shaping their reaction to the rice theft.

The theft stemmed from the director of Our Future's own resentment of ex-combatants, which I will address fully in the next section. 'He told us he was impounding the rice,' stated a PADO board member, 'because ex-combatants should not be directing this development project.' The ex-combatant described the accusations levelled at PADO when he returned to Malempa with raffia bags into which the rice was to be divided.

> The women were crying and wringing their hands, they were desperate, that the rice had been taken. They demanded to know where we had put it and that we should bring it back. I told them that we did not impound the rice and they did not believe me. They accused me of tricking them by going to town for the bags and sending Our Future to take it for us. They said it was a way for me to be away when I had these men steal the rice. They did not trust me, even when I explained that I was away because I was looking for a way to release money from the account to purchase the bags, because Our Future did not release it for us.

The fact that Malempa residents accused PADO of stealing the rice illustrates how long-simmering resentment of ex-combatants coloured their interpretation of the theft. The initial collective reaction was that Our Future was transporting the rice as 'implementing partner' and that PADO had been making the decisions. The critical points of friction were that PADO board members were controlling, they wanted ownership of the project, treated residents as free labour and were still 'hungry' for surplus rice for their own purposes. Despite the fact that it was PADO, and not Our Future, that had created and found funding for the project, in the eyes of Malempa residents, they bore the greatest responsibility because the project was simply another illustration of their arrogance and unwillingness to reintegrate humbly, on civilians' own terms. In essence, PADO was following the letter of the project, and their failure to engage with civilian farmers on deeper levels of redress and reconciliation meant they shouldered the blame for project failure, even as civilians accepted the project because they needed funding for agriculture. For ex-combatants to treat civilians as a source of income was unacceptable. Thus agricultural funding being available only for cooperative ventures was revealed as untenable for accomplishing agriculture and reconciliation. In practice, because agriculture was about self-determination for all involved parties, it accomplished neither.

Despite several years of cooperative efforts between PADO and Malempa, the judgement of ex-combatant guilt in the theft was immediate and visceral. The fact that civilians leapt to unsubstantiated accusations, even with evidence that Our Future was responsible, revealed their deep-seated unhappiness with having to partner with and essentially work for ex-combatants. That PADO had softened their stance towards the project by allocating seed earmarked for the organization from the communal farm to the farmers themselves could not undo the damage that had already occurred. To resent ex-combatants was to take a moral stance against the war, its perpetrators, their lavish reintegration packages and their refusal to reintegrate humbly as equal participants in a post-war world. In addition, civilians aired their resentment of enforced cooperation and their consistent inability to determine the course of their own food production, and hence their own lives. Our Future, as an NGO managed and staffed by civilians whose stated purpose was to promote post-war development, was initially above reproach.

The Director of Our Future Lashes Out at Post-war Aid through His Organization

Why was the rice stolen? The fact that the director stood to gain financially notwithstanding, deeper ideological issues were at stake. A different ex-combatant CBO in the chiefdom had paired with Our Future on a project at the same time, which ended in failure because the seed disappeared after the fields had been ploughed. According to one participant, 'He [the director] will take from these projects because he wants them to fail.' Though I did not have access to the director of Our Future, the board members of PADO documented in a letter to him on 3 June 2008, which they copied to 13 different officials in Sierra Leone and the donor community, that he admitted that he was driven by his own resentment of ex-combatants. His goal in stealing the rice was to consolidate community opposition to ex-combatant CBOs, in essence taking a destructive stand against enforced cooperation itself.

In the letter, the PADO board noted that they did not accept the fact that he was seeking revenge on them for being ex-combatants. They paraphrased what he had said: 'You told us you have come to destroy this project because you want revenge on the people who brought war to this country, and you have hate in your heart because those who have destroyed this country now want to enjoy it.' As the director was on the scene for the theft, it appeared that he cared little for Our Future's legitimacy and funding once accusations were lodged; the proceeds from the stolen rice and the satisfaction of damaging the project apparently outweighed these concerns. Indeed, the PADO board had no power to influence the police or government officials, who ignored their pleas for intervention and investigation.

The PADO board accused the director specifically of sabotaging peace rather than agriculture. Using the language of reconciliation, they disingenuously appealed through their documentation to the donor's goals for the project, rather than their own. They accused the director of failures of reconciliation because he had not organized official activities. They defended themselves as ex-combatants and attempted to chide the director for his actions by appealing to the national call for forgiveness that had occurred during the peace process:

> The behaviour you have displayed has had the effect of dividing the community from us. Now there are lots of conflicts and confusion among members of the community. Instead of building peace, you have nurtured conflict. Presently because of your behaviour we the executive members of PADO cannot feel secure because the community blame us for directing you. Your behaviour was terrible and seemed to be directed at undermining the peace.

Ex-combatants had followed the letter of the law and the director had not, despite the fact that Malempa residents themselves asked PADO to rectify the inborn inequalities of the project. As their CBO was an official reintegration activity, PADO members resorted to reintegration language to defend themselves to the donor by framing the director as a warmonger.

PADO board members struggled to find civilian or official support for an investigation into the theft, which might have exonerated them from blame for taking the rice. PADO sent letters asking for an investigation to the Paramount Chief, members of parliament, the mayor of the city in which Our Future was headquartered and the police. Their requests went unanswered and no charges were filed. One board member wrote a newspaper editorial advertising Our Future's corruption but the NGO was not suspended. They halted their efforts after a final plea to the Anti-Corruption Commission failed to result in an investigation.

PADO struggled to maintain itself as an organization. According to one board member, 'We were just so lethargic for a long time, we could not muster the energy to attempt another project.' One member left for Freetown in 2007 and never returned. The other three attempted to heal the rift with Malempa when they were offered the leftovers of an agricultural project that a different NGO was undertaking in 2009. They were given fifty bundles of cassava sticks to plant, and, 'with joy in our hearts', according to one, returned to Malempa to propose another project. They deposited the sticks in the centre of town after the chief assured them that though the women were not around to collect them – many were working on their individual farm plots a mile from the village – they would gather to divide the sticks for planting that evening. Once again, PADO was delivering a project and Malempa residents were expected to provide the labour for it.

When PADO board members returned to assess progress two weeks later, they found the stick bundles exactly where they had left them, shrivelled and dead.

The chief was evasive when questioned, alluding to the fact that the women did not want to work with PADO any more because they did not know when the fruits of their labour would be taken from them once again. Though three years had passed since the initial theft and Our Future was not involved, the women were too demoralized and suspicious to start a new project with PADO. They refused to plant despite the fact that cassava, unlike rice, is a 'theft-proof' crop because the tubers can be left in the ground until needed for consumption. Again, food production was stymied by the inequality of the terms of cooperation and the unaddressed issues of reintegration.

Lessons Learned: Can Cooperative Agriculture Promote Reintegration and Food Security?

The failure of PADO's projects – and the potential pitfalls of other agriculture-as-reconciliation projects – highlights several key questions that must be answered before cooperative agriculture becomes an effective tool of post-war reintegration and potentially of food security:

- Is cooperation being forced on ex-combatants and civilians because donor funding is available only under the auspices of reconciliation?
- What is the relative status of the various participants in the project, and who has the right to determine the provision of budget, labour and even wages?
- What is the status of ex-combatant reintegration when projects begin? Do civilians approve of the reintegration process? Do they feel pressured to accept ex-combatants when they are not ready?

In this case study, the pressure to earn an income as 'strangers' with no marketable skills and no other access to land moved PADO to treat the project as their business rather than a route to reintegration, even as they hid behind the language of reintegration when the project fell apart. Because PADO knew that Malempa needed ex-combatants to register as a CBO, PADO claimed ownership, 'giving' the project to Malempa, which increased civilian resentment of ex-combatants. Malempa farmers and the director of Our Future had unaired grievances with DDR, ex-combatants and cooperative agriculture that resulted in arguing about money and eventually outright theft. In hindsight, it seems coercive to delineate an essential life-giving activity only as a means for reconciliation if no other avenues are presented for people to fund their own agricultural activity, or to address – or not – their issues with ex-combatants. If cooperative agriculture is used as a mechanism for reconciliation, this must be done in parallel with other projects focused specifically on food production in war-affected villages, with no reconciliation caveats. Reconciliation within agricultural projects is likely to fail if it sidelines or undermines the possibility of self-determination.

References

Bolten, C. 2008. 'The Place Is So Backward': Durable Morality and Creative Development in Northern Sierra Leone. Doctoral dissertation submitted to the University of Michigan. Ann Arbor: ProQuest.

Chayes, A. and Minow, M. 2003. Introduction, in *Imagine Co-existence: Restoring Humanity after Violent Ethnic Conflict*, edited by A. Chayes and M. Minow. New York: Jossey-Bass, 1–23.

Fanthorpe, R. 2001. Neither citizen nor subject? 'Lumpen' agency and the legacy of native administrations in Sierra Leone. *African Affairs*, 100, 363–86.

Ginifer, J. 2003. Reintegration of ex-combatants, in *Peacekeeping in Sierra Leone: UNAMSIL Hits the Home Straight*, edited by M. Malan et al. Pretoria: Institute for Security Studies, 39–52.

Murphy, J. 2003. *Getting Even: Forgiveness and Its Limits*. New York: Oxford University Press.

Solomon, R. 1983. *The Passions: The Myth and Nature of Human Emotions*. Notre Dame: University of Notre Dame Press.

Utas, M. 2005. Building a future? The reintegration and remarginalization of youth in Liberia, in *No Peace, No War: An Anthropology of Contemporary Armed Conflict*, edited by P. Richards. Athens, OH: Ohio University Press.

PART IV
Conclusion

Chapter 18

Concluding Remarks: Looking to the Future – Agriculture Post-conflict

Rebecca Roberts and Julia Wright

This book aims to initiate a discussion about the immediate and long-term impact of conflict on agriculture, some of the issues that have to be considered for establishing sustainable agriculture and some of the grassroots and externally supported approaches to rehabilitate the agricultural sector. There are still many areas for research, and many areas that have not been discussed here, for example, the difficulties in finding a balance between large-scale internationally driven assistance, demanding the development of national policies such as poverty reduction strategy papers, and small-scale community-led processes.

To date, there has been little debate on the nature of agricultural strategies in post-conflict situations. For international non-governmental organizations (INGOs), the strategy for dealing with post-conflict situations tends to draw directly from the strategy for agricultural rehabilitation post-natural disaster. However, the post-conflict context is characterized by a much higher degree of personal insecurity, whereas natural disasters tend to more directly impact the productive base through flooding, landslides or drought. Despite the differences, there are similarities between natural and man-made post-disaster scenarios, and these include the levels of displaced persons, the breakdown of traditional structures and processes, the loss of life, the destruction of the physical infrastructure, and the increased threat of disease and food insecurity. Interventions are, in theory, guided by the technical standards of the *Sphere Handbook* (2011): the Humanitarian Charter and Minimum Standards in Humanitarian Response. This is the chief reference book for humanitarian aid in general, and it addresses agriculture in a small section on Primary Production within its minimum standards for food security and nutrition (2011: 204). This section recommends actions and guidance notes that are generally supportive of a holistic approach to agriculture. Nevertheless, approaches to agriculture programmes vary, and to a large extent depend on the knowledge and understanding of the implementing body.

Owing to the lack of preparation and the complexity relating to emergency operations, many relief interventions have been unsuccessful in restoring the full agricultural capacity of farming systems (Bushamuka 1998, Hines, Wickrema and Van Straaten 1998). In the process of moving from relief to development, there has been a tendency to commence with food aid and then to phase this out and replace it with the supply of seeds, tools and fertilizers. This has been undertaken without due regard to the Sphere recommendations. Further, there has been a disregard for the fact that farmers require access to a broad spectrum of inputs and services that are interdependent, such as water, fertilizers, credit and market information services, and that the process is a participatory one (Longley et al. 2007).

In conflict zones and areas emerging from conflict, military units are playing a greater role in the reconstruction of agricultural systems, and particularly so in regions where, owing to insecurity, the risks are too great for civilian staff to operate. Priorities here are to provide the local population with access to a secure food supply as a first step toward stabilizing regions, and then to improve the livelihoods of communities through agricultural development as a longer-term means to ensure stability (Peck and Kemmet 2011). To do this, military units develop a broad strategy, and draw on knowledge and expertise from agricultural experts internationally and/or nationally. The agricultural approach taken, therefore, is based partly on the military perspective and understanding of agricultural issues in general and in their areas of operation, and partly on the perspective of these experts. These interventions have a limited life span and so activities are selected that will show impacts in a relatively short time over a wide area or large number of households. The British government's Stabilisation Unit provides one example. It promotes basic guidance for food and livelihoods in its handbook on Quick Impact Projects (undated). In terms of food, the military may be involved in providing logistics for transporting food supplies to civilian populations. Livelihood support is seen as crucial for the reintegration of former combatants and displaced persons, and so intensive vocational training and labour-intensive activities are highlighted. No mention is made of agriculture, but the most appropriate technical standards are referred to as contained in the *Sphere Handbook*.

Overall, the *Sphere Handbook* contains sound recommendations for holistic support to agriculture. Yet these require a large degree of participatory input by local populations, and time for evaluations and assessments, all of which may not be feasible or funded in a post-conflict situation. Further, these recommendations are contrary to the conventional agricultural approach that is more familiar to the military and private sector. In fact, Longley et al. (2007) argue that the objective of agricultural support to fragile states emerging from conflict has been a more conventional focus on increasing production per se, whereas it should aim to enhance consumption, markets and livelihoods, and to do so requires increased access by farmers to appropriate technology options, inputs and services. In this sense, and whatever the theoretical guidelines, the paucity of evidence suggests that agriculture in post-conflict situations tends to mirror agricultural approaches elsewhere and is limited by the degree of knowledge of the implementing agency. Further, although guidelines recommend the use of sustainable agricultural techniques, these can be applied piecemeal and without a clear, holistic methodology or design framework.

It is apparent then that more research is needed to understand fully how to support the agriculture sector during and after conflict. The lessons from the case studies here show that there are many potential approaches to supporting agriculture post-conflict, but that they need to be tailored to each specific context. They need to take into account local capacities and traditional practices, as well as the potential impacts of the conflict, which may mean that it is not possible or desirable to return to pre-conflict agricultural practices. The disruption to education, rule of law, land rights and access to natural resources needs to be tackled in the post-conflict period to ensure that the population has adequate skills, resources and rights to (re-)establish the agricultural sector. To maximize the contribution of agriculture to the economy there need to be coherent national strategies based on an accurate understanding of the state of the country. In post-conflict environments, obtaining even basic facts about the population size and its distribution can be difficult. In addition, where appropriate, ways of accessing the international markets so that agricultural produce can be exported need to be explored. This would precipitate a move away from subsistence agriculture. Investment in agricultural processing is also necessary so that countries can add value to exported agricultural goods. At the same time, institutions to support and develop the agricultural sector and research institutions to strengthen agricultural practice should be established.

In countries emerging from conflict, the modernization and development of the agriculture sector to improve its economic potential and increase competiveness on the international market needs to be approached with caution. Large-scale international interventions can undermine the economies of the people they are trying to help unless they are managed well by national authorities. Satisfying donor conditionality and fulfilling the expansive terms of international policies can be a drain on national resources and a distraction from the key concerns of rehabilitating the agriculture sector and attaining sustainable food security. What is presented as food aid can be a form of 'dumping' as developed countries attempt to relieve themselves of their surplus production. Such 'gifts' skew the local economy and discourage local production. Growing produce to appeal to the economically lucrative Western markets can result in a loss of traditional farming practices and crops which has a negative effect on the long-term food security of populations recovering from conflict.

It is imperative that there is food security at the community level to ensure survival. This means that external organizations, wherever possible, should continue to support agriculture during conflict. However, it is also important to recognize that communities have their own traditional coping mechanisms and abilities to maintain agricultural production. The capacity to continue traditional farming practices provides important psychosocial support. If external support becomes available it should aim to build on these coping mechanisms and avoid establishing parallel systems that may undermine existing capacities. The central importance of food security means that, particularly in post-conflict environments where social dynamics are damaged and complex, incorporating agricultural programmes into broader peacebuilding initiatives may be inappropriate. A society divided by conflict is unlikely to be united by an agricultural project alone. Food security is needed to avoid a return to conflict and this must be prioritized over other objectives. However, there is evidence that externally supported interventions to promote agricultural production can have wide-reaching positive effects. First, agriculture provides a legitimate entry point for external actors into conflict-affected communities and justifies the presence of external actors during ongoing conflict. Second, participatory approaches help to establish relationships between external actors and communities as well as within communities, and strengthen the role of civil society organizations. A concentrated effort on a common concern such as food production helps to build and reinforce informal social networks. Such impacts are particularly important among conflict-affected populations where trust and social bonds are likely to have been damaged. Finally therefore, it seems that agriculture can help to promote reconciliation among conflict-affected people and assist individuals in recovering from the trauma of conflict. It is important to note though that these positive benefits are by-products of programmes explicitly designed to promote agriculture, and not the outcome of a peacebuilding programme using agriculture as a tool for reconciliation.

In addition to learning from past experiences of promoting agriculture during and post-conflict, further research is needed to mitigate against potential future conflicts. With climate change and an increasing global population, the possibilities for conflict over what may become scarce resources are clear. It is evident, given the number of famines and cases of malnutrition and starvation, that even without the threat posed by climate change we are unable to feed the world's current population. Active measures need to be taken to manage resources fairly and efficiently to reduce the risk of conflict and ensure food security for all.

The case studies in this book demonstrate that there are many examples of best practice and lessons learned upon which to begin exploring the potential to rehabilitate agriculture post-conflict. However, to date, the challenges of recovering the agricultural sector after conflict have been overlooked. Given that food security is essential to the survival of the world's population and that equitable access to resources and adequate food help are needed to avoid a return to conflict, practitioners, policymakers and academics need to devote more attention to establishing sustainable agricultural production. The challenges of climate change and a growing population increase the potential for conflict over resources so it is vital that we develop both effective mechanisms to manage resources and practical sustainable approaches to ensure agriculture recovers after conflict.

References

Bushamuka, V.N. 1998. Restoration of Seed Systems and Plant Genetic Resources After Disasters: A Synthesis of Background Papers. Ithaca: Cornell University.
Hines, D., Wickrema, S. and Van Straaten L. 1998. Food and Seed Assistance in the Recovery from Crisis. Background paper for the International Workshop on Developing Institutional Agreements and Capacity to Assist Farmers in Disaster Situations, Rome, 3–5 November 1998.

Longley, C., Christoplos, I., Slaymaker, T. and Meseka, S. 2007. Rural Recovery in Fragile States: Agricultural Support in Countries Emerging from Conflict. Natural Resource Perspectives 105. London: Overseas Development Institute.

Peck, E. and Kemmet, L. 2011. Application of a network science approach to post-conflict/post-disaster agricultural system reconstruction and development. Selected paper prepared for presentation at the Agricultural & Applied Economics Association's AAEA & NAREA Joint Annual Meeting, Pittsburgh, PA, 24–26 July 2011.

Sphere Project. 2011. The Sphere Handbook: Humanitarian Charter and Minimum Standards in Humanitarian Response. Rugby: Practical Action Publishing.

Stabilisation Unit. Undated. Quick Impact Projects Handbook. London: Stabilisation Unit [Online]. Available at: http://www.stabilisationunit.gov.uk/attachments/article/520/QIPs%20handbook[1].pdf [accessed: August 2011].

Index

References to figures are shown in *italics*. References to tables are shown in **bold**. References to notes consist of the page number followed by the letter 'n' followed by the number of the note, e.g. 194n4 refers to footnote no.4 on page 194.